VOLCANOES OF THE CASCADES

THEIR RISE AND THEIR RISKS

RICHARD L. HILL

GUILFORD, CONNECTICUT
HELENA, MONTANA
AN IMPRINT OF THE GLOBE PEQUOT PRESS

Illustrations by Steve Cowden

Text design by Linda Loiewski

Library of Congress Cataloging-in-Publication Data

Hill, Richard L.
Volcanoes of the Cascades: their rise and their risks / Richard L. Hill.—1st ed. p. cm.—(A Falcon Guide)
ISBN 0-7627-3072-2
1. Volcanoes—Cascade Range. 2. Geology—Cascade Range. 3. Cascade Range. I. Title. II. Series.

QE535.2.U6H55 2004
551.21'09795—dc22 2004047157

ISBN 978-0-7627-3072-8
Manufactured in China
First Edition/Second Printing

CONTENTS

THE CASCADES

For airline passengers flying in from the east over the Pacific Northwest, the snowcapped mountains that seem to spring up outside their windows provide a refreshing transformation from the stark expanses of the Great Plains.

Stretched below are the majestic Cascade Mountains, a relatively young volcanic range that extends from British Columbia to northern California in a nearly 700-mile arc. The centerpiece of the Northwest's renowned beauty, the Cascades run almost parallel to the Pacific Ocean some 100 to 150 miles to the west.

These rugged, picturesque peaks draw hundreds of thousands of visitors to their flanks each summer, and several serve as the focal points for national parks and monuments, including Mount Rainier, Mount St. Helens, Crater Lake, Newberry Volcano, and Lassen Peak. The Cascade Range also plays a key role in the Northwest's climate. Moisture-bearing winds and winter storms from the Pacific Ocean are intercepted by the mountains, making the western halves of Washington and Oregon moist and lushly forested and their eastern portions arid and desertlike. The Cascades have had a significant economic impact on the Northwest. Their thick forests have been an important source of jobs and revenue, and their snow-covered slopes provide water for hydropower and agricultural irrigation. The mountains, along with their lakes and rivers, also serve as a recreation attraction for residents and tourists who enjoy camping, hiking, fishing, whitewater rafting, and skiing.

The Cascades' thirteen major volcanic centers may appear to passersby on Interstate 5 or in the air to be isolated peaks that bear little relationship to

Glacier-covered Mount Rainier is the king of the Cascade Range.

each other. But they are linked together by a tectonic-plate process that keeps a few of the combustible mountains popping. Though the larger peaks like Mount Rainier near Seattle and Mount Hood near Portland grab the most attention, the Cascades are made up of thousands of smaller extinct volcanoes that lived short, explosive lives.

Of the landscape-dominating peaks, Mount Rainier is the tallest, towering over Washington's Puget Sound area at 14,410 feet. Mount Shasta in remote north-central California follows at 14,162 feet. Mount St. Helens in southwest Washington became the shortest significant volcano at 8,364 feet after its infamous 1980 eruption ripped 1,300 feet from its summit.

Underlying their elegance, however, is the disquieting fact that the range's thirteen major peaks are potentially hazardous volcanoes. Seven have erupted in the past 250 years, with huge explosions in the past century at Lassen Peak and Mount St. Helens leaving onlookers in awe.

Along with Alaska's Aleutian Islands, the Cascades are an active part of the "Pacific Ring of Fire," a rough circle of volcanic ranges found along most of the Pacific Ocean's rim. About three-quarters of the earth's active and dormant volcanoes above sea level are in this ring, with nearly 500 of them having

WHO NAMED
THE **CASCADES?**

Who named the Northwest's major mountain range the "Cascades"? It's a whodunit that may never be resolved.

A logical assumption is that the mountains were named after the famous cascades of the Columbia River, which flows between present-day Washington and Oregon. Before twentieth-century dams, the cascades east of Portland in the Columbia River Gorge were dangerous rapids feared by explorers and early settlers.

In their authoritative book *Oregon Geographic Names,* the late Lewis A. McArthur and his son, Lewis L. McArthur, explain that the Klamath Indians are the only known tribe that had a name for at least a portion of the Cascade Range—yamakiasham yaina, meaning "mountains of the northern people."

A crude map by Manual Quimper, a Spaniard, first labeled the range as "Sierras Nevadas de S. Antonio" in 1790. He was followed two years later by English explorer George Vancouver, who named a few of the volcanoes after Royal Navy admirals (Rainier and Hood, for example) but only referred to the range as "snowy range" and "ridge of snowy mountains." Lewis and Clark stuck to the snow theme in their journals, with such references as "the Western mountains covered with snow."

The McArthurs say it is possible that the famed botanist David Douglas—the Douglas fir is named for him—may have named the mountains. He refers repeatedly in his 1820s journals to the "Cascade Mountains" or "Cascade Range of Mountains," but Douglas never claims to have originated the name, so the credit may lie elsewhere.

In the 1830s an enthusiastic fan of the Oregon country, Hall J. Kelley, named the chain the "Presidents' Range" and labeled individual mountains after U.S. presidents. But his moniker never stuck. Lt. Charles Wilkes, a surveyor with the U.S. Navy who led a four-year expedition that circumnavigated the globe, identified the mountains as the Cascade Range in 1841. They've been the Cascades ever since.

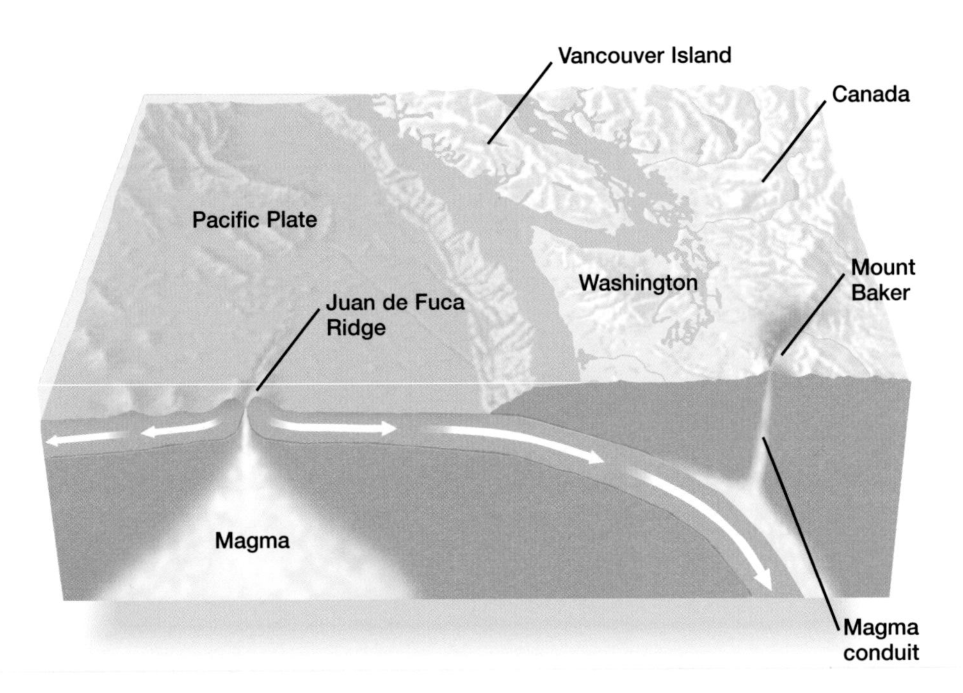

The eastward-moving Juan de Fuca Plate, which forms at the seafloor Juan de Fuca Ridge, plunges beneath the North American Plate. As the denser oceanic plate is driven downward, it encounters high pressure and temperature that partially melt the solid rock at the base of the upper plate. The newly formed magma rises through cracks to the surface, where it erupts, forming the chain of volcanoes known as the Cascades.

erupted in recorded history. In addition to rumbling volcanoes, the Ring of Fire also shakes with incessant earthquakes, generated by the same tectonic-plate processes wherein the huge Pacific Plate and several smaller plates collide with and slide beneath other plates. This volcano- and earthquake-producing process is known as "subduction."

The Cascade volcanoes' origins lie along one of these subduction zones. About 200 miles off the Northwest coast rises the seafloor volcanic chain known as the Juan de Fuca Ridge, where molten rock, called "magma," oozes onto the seafloor to form new oceanic crust. This fresh crust adds to the Juan de Fuca Plate, which moves eastward toward land at a rate of about 2 inches each year. The young, denser plate collides with and is plunged—subducted—beneath the North American Plate between 60 and 150 miles offshore. Scientists say powerful earthquakes occur on average about every 500 to 600 hundred years where the two plates meet. The last one was about 300 years ago. We could be due.

As the Juan de Fuca Plate dives deeper and deeper beneath the continent,

THE EARTHQUAKE **RISK**

The same tectonic-plate process that feeds into the Cascades' major volcanoes also causes earthquakes—a real threat to the nearly ten million people who live west of the mountain range. A huge earthquake could result in the nation's largest natural disaster.

The 600-mile-long area where the Juan de Fuca Plate plunges beneath the North American Plate—about 60 to 150 miles offshore—is known as the Cascadia Subduction Zone. In this area the plates are locking up as they slide past each other, building enormous strain that is released by huge magnitude 8 and 9 earthquakes.

Scientists have found evidence that these earthquakes—and the resulting huge waves called "tsunamis" have repeatedly rocked the Northwest coast from Vancouver Island to northern California for the past several thousand years. Such events undoubtedly will strike again.

Similar subduction zones caused the planet's two largest recorded earthquakes: a magnitude 9.5 quake on the coast of Chile in 1960 and a magnitude 9.2 quake in southern Alaska in 1964.

Back in the Pacific Northwest, as the plates shove against each other, the upper North American Plate near the coast slowly is being pushed upward and eastward. When the accumulated pressure is released in a powerful earthquake, the coastline snaps back—plunging by as much as 6 feet. Although these so-called megathrust subduction-zone quakes, which can last for three to five minutes, have not occurred in Oregon and Washington's brief recorded history, signs of abruptly buried marshes and forests along the coast are indications of such quakes. Tsunami records in Japan, tree-ring dating, and other evidence indicate that the last such quake was a magnitude 9 in January 1700. Recent studies show that at least a dozen earthquakes have hammered the Northwest coastline in the past 6,700 years. No one knows when the next one will occur, but concerned scientists have spread the word that the public should prepare for such a quake.

A future quake could devastate the coastline as well as cause landslides on the steep slopes of some of the Cascade volcanoes. Landslides on Mount Rainier, for example, could develop into fast-moving mudflows down its river valleys to the large developed communities currently living peacefully on its flanks.

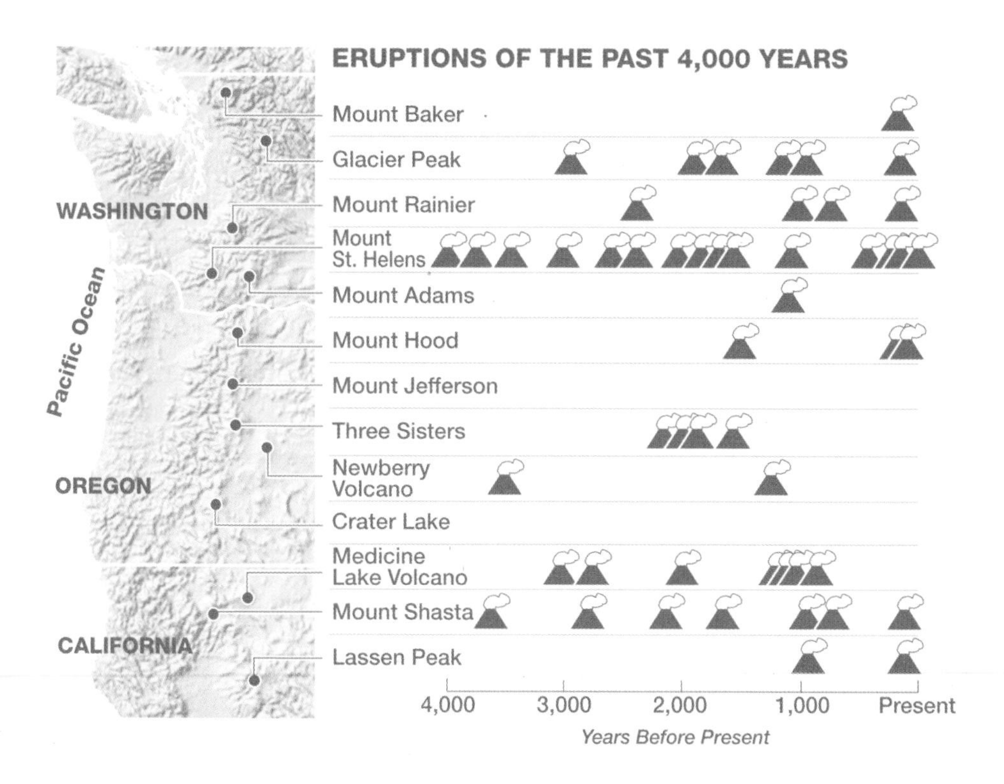

it gets hotter and more compressed. Eventually, about 60 miles below the surface, the plate encounters intense pressure and temperature. Water from the plate is drawn out in a process that helps melt the base of the overlying North American Plate. These blobs of hot, fresh magma rise through cracks and into chambers a few scant miles below the earth's surface. Increasing pressure eventually forces the gas-filled magma upward to the surface, where it erupts—often violently—to form or add to a volcano. In addition to tectonic plate processes, heavy rainfall also appears to play a key role in the Cascade Range's geologic structure. Scientists reported in a study in 2003 that the abundant precipitation not only leads to a high erosion rate, but causes bedrock to be pulled toward the earth's surface faster in some places than others. Researchers at Yale, the University of Washington, and the University of Michigan said that the uplift of rock on the Cascades' west flank, where precipitation is high, is three times faster than elsewhere in the range.

Because of the way they were formed, most of the major Cascade peaks are classified as "composite" volcanoes, also known as "stratovolcanoes." They are what most people picture when they think of a volcano: steep-sided mountains with sharp, pointed tops and a vent at the peak.

Stratovolcanoes are built of alternating layers of lava and volcanic debris. They tend to explode when they erupt, causing a lot of damage. A good example is the 1980 blast of Mount St. Helens, which killed fifty-seven people and devastated the nearby landscape. In contrast are the gently sloping "shield" volcanoes, such as those in Hawaii, with their low profiles and lava that is less gooey than that of stratovolcanoes and can flow for up to hundreds of miles. Because they tend not to be explosive, shield volcanoes seldom cause injury or death, though they may result in extensive property damage. Two of the volcanic centers in the Cascades—Newberry Volcano in central Oregon and Medicine Lake in northern California—are primarily shield volcanoes.

The stately Cascade peaks that tower above the landscape are newcomers to their ancient Northwest neighborhood that originated as long as 300 million years ago. The modern range as we see it today began forming about five to seven million years ago, and its major peaks have risen within the past one and a half million years atop volcanic rock that spewed to the surface during the past thirty-six million years.

Mount St. Helens and Mount Baker are the youngest of the principal volcanoes, still infants at only 30,000 to 40,000 years old. Their older relatives, such as Mount Rainier and Mount Hood, range up to three-quarters of a million years old. Most of the thirteen primary Cascade peaks have exploded in the past 4,000 years, with seven having erupted within the past 250 years. All are presently dormant, meaning that although they may appear quiet, they are still active volcanoes that can be expected to erupt in the future. Even if they do not awaken, however, the volcanoes still pose risks to those who live, work, and play on their slopes and river valleys.

For more information:

USGS Cascades Volcano Observatory

1300 Southeast Cardinal Court, Building 10, Suite 100
Vancouver, WA 98683-9589
(360) 993–8900
vulcan.wr.usgs.gov

Global Volcanism Program

Department of Mineral Sciences,
National Museum of Natural History, Room E-421, MRC 0119, P.O. Box 37012,
Smithsonian Institution, Washington, D.C. 20013
www.volcano.si.edu/gvp

VOLCANIC HAZARDS

The glaciers and wildflowers that cover the Cascade Mountains provide a scenic disguise for the potentially lethal forces that lie within. As the spectacular volcanoes typically stay asleep for centuries, those who visit or live in the Northwest often pay little heed to their latent destructive power.

Those nonchalant attitudes changed within a few seconds on May 18, 1980, when the eruption of Mount St. Helens served as a shocking reminder of a volcano's deadly power. The explosion caused more than $1 billion in damage and killed fifty-seven people, the largest number of U.S. fatalities from a volcanic blast in recorded history.

Despite how popular movies depict eruptions, the biggest threats from volcanoes aren't actually lava flows. Volcanoes—including those in the Cascades—can kill in a remarkable number of ways, from ashfall to mudflows and avalanches. The steep ice and snow-covered peaks can be deadly during an eruption, when magma and other material are ejected onto the volcano's surface and into the air. But dangers can exist even without an eruption.

Here's a menu of what these dangerous volcanoes can dish out:

Pyroclastic flows. Dense, high-speed avalanches of hot gas, ash, and rock fragments that race down the slopes during an eruption. These flows can reach speeds of up to 150 miles per hour and temperatures of 1,500 degrees Fahrenheit. Pyroclastic flows have been the dominant killer from eruptions worldwide, killing nearly 80,000 people in the past half-dozen centuries.

RATING

THE **RISKS**

As a geophysics researcher at the University of Washington, Dr. Steve Malone has been studying earthquakes near and under Cascade volcanoes for more than three decades. He has been a key leader in setting up a network of earthquake sensors throughout Oregon and Washington, including on many of the region's restless mountains.

Though the Cascade peaks are infamous for their explosive volcanic eruptions, earthquakes—even moderate ones—also pose a threat to people and property. Earthquakes can trigger flows of mud, rock, water, and ice from the steep-sided volcanoes.

In recent years Malone has rated the volcanoes according to the hazards that earthquakes pose and how likely they are to erupt in our lifetimes. Malone rated Mount Rainier as posing the biggest volcanic earthquake threat to people because of the 14,410-foot peak's steep snow- and ice-covered slopes and its proximity to the urban areas in the Puget Sound area. Based on seismicity, the geologic record, and slope stability, Malone followed Rainier with Mount Hood and Mount Shasta.

As for eruptions, Malone rated Mount St. Helens as most likely to erupt in our time—again. His list of those moderately likely to erupt in our time includes Mount Rainier, Mount Hood, Mount Shasta, and Lassen Peak. And those he terms as "not *as* likely to erupt in our time": Glacier Peak, Mount Jefferson, Three Sisters, Mount Baker, and Mount Adams.

Pyroclastic surges. Less dense than pyroclastic flows, surges can be even more disastrous as they travel at speeds of up to 300 mph. Unlike flows, surges are not confined to valleys and other topographic features but can easily move over ridges, hills, and other obstacles. Scientists with the U.S. Geological Survey estimate the pyroclastic surge produced by the eruption of Mount St. Helens accelerated to nearly 300 mph before slowing and spreading out from the volcano. When the surge slowed, several people were able to escape only by driving at speeds up to 100 mph.

◄ A pyroclastic flow speeds down Mount St. Helens in August 1980.

Photographer Reid Blackburn's car was buried in the 1980 eruption of Mount St. Helens. Blackburn died in the blast.

Lahars. An Indonesian word for a flow of mud, rock, and water that behaves like a thick soup of wet concrete as it speeds down a volcano at 15 to 40 mph. Also called mudflows or debris flows, lahars can destroy everything in their path and are regarded as the second most dangerous volcanic hazard. A single flow from the Nevado del Ruiz volcano in Colombia killed more than 23,000 people in 1985. Cascade volcanoes are notorious producers of lahars, which can be triggered either by large landslides or by the rapid melting of snow and ice from even small emissions of heat. Lahars follow river valleys and are a potential hazard to communities downstream from glacier-cloaked volcanoes, such as Mount Hood and Mount Rainier.

Landslides. Rapid downhill movements of rock, snow, ice, and other material. Also called debris avalanches, landslides often occur when steep upper areas collapse under the force of gravity. In the Cascade volcanoes, warm acidic water circulates in cracks and porous areas inside the mountain, changing strong rock to slippery clay. Magma pushing up into the volcano, small earthquakes, and even a heavy rain can trigger avalanches. The eruption of Mount St. Helens was unleashed by the largest avalanche in recorded history. Loosened by a 5.1-magnitude quake about a mile under the volcano, the upper north flank—weakened and deformed by two months of eruptive activity—collapsed and nearly three-fourths of a cubic mile of rock, mud, ice, snow, and water plunged away at speeds of up to 180 mph.

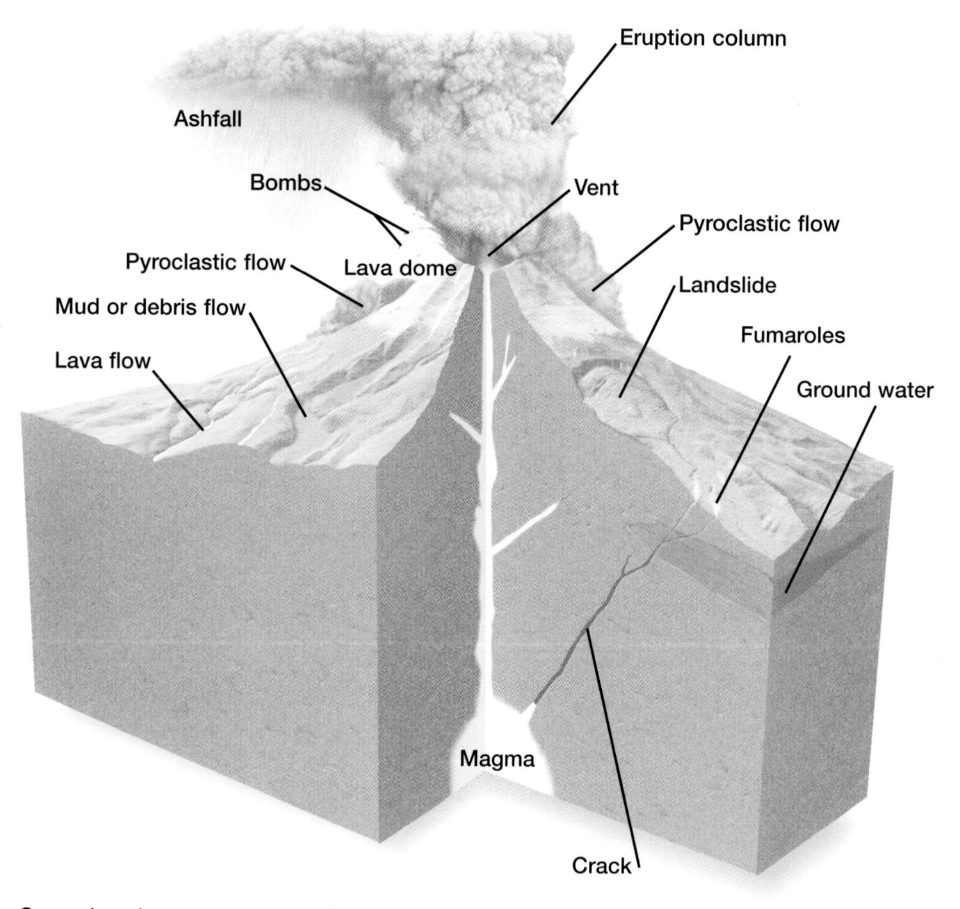

Cascade volcanoes pose a variety of hazards to people and property, from ash-filled clouds that can damage aircraft engines, to fast-moving lahars and debris avalanches that can sweep away forests and fill rivers.

Lava flows. Hot streams that destroy anything in their path but usually flow slowly enough that people and animals can easily move out of the way. Molten rock that rises from beneath a volcano is known as magma, but after it erupts from a volcano, it is called lava. Lava flows can cause forest fires and block or alter the course of rivers. Sometimes mounds of thick, pasty lava pile up at volcanic vents during eruptions, forming what are called lava domes. The lava dome created in the crater at Mount St. Helens grew 876 feet high and nearly 3,500 feet wide. Lava domes can be dangerous. For example, when hot, new ones on Mount Hood have became unstable and collapsed, avalanches have swept down the mountain. Such an event hasn't occurred in recent history but could happen again.

The lava dome in Mount St. Helens rises 876 feet above the crater floor.

Tephra. A mix of material—including fragments of volcanic rock and glass ranging in size from ash to larger blocks, or "bombs"—that is blasted into the air or carried upward by hot gases in huge plumes or columns. Clouds of volcanic ash can be carried hundreds of miles, disrupting air traffic, causing buildings to collapse, and damaging machinery, electronic equipment, and crops. Heavy ashfall also can injure fish and kill vegetation along rivers and streams.

Gases and acid rain. Sulfur dioxide, carbon dioxide, hydrogen sulfide, hydrogen chloride, hydrogen fluoride, and other toxic compounds emitted by volcanoes along with steam from heated groundwater. These gases can kill people and livestock, damage crops, and contaminate water supplies.

Because of the array of dangers posed by volcanoes, scientists know they need to keep a close eye on the Cascades. When Mount St. Helens erupted in 1980, Congress provided funding for the U.S. Geological Survey to

A VOLCANO

SWAT **TEAM**

Restless volcanoes throughout the world keep a group of scientists based at the U.S. Geological Survey's Cascades Volcano Observatory (CVO) in Vancouver, Washington, on the go. Members of the Volcano Disaster Assistance Program stay busy helping to monitor active volcanoes in countries that ask for expertise and equipment. They've been called to check out volcanoes in Central and South America, the South Pacific, Asia, Africa, and the Caribbean.

The world's only mobile response team was established in 1986 as a result of the 1980 Mount St. Helens eruption and the devastating 1985 eruption of Colombia's Nevado del Ruiz volcano, which killed more than 23,000 people with the huge fast-moving flows of mud and debris called lahars.

Mount St. Helens proved to be a perfect laboratory for volcano scientists. More than two dozen eruptions followed the blast that killed fifty-seven people. These eruptions provided researchers at the newly established CVO the chance to develop eruption-forecasting tools and skills. At a country's invitation, the team can rapidly set up instruments to collect and analyze information on a fidgety volcano's activities, such as earthquakes, ground deformation, gas emissions, and lahars. With that data, scientists can give a rapid and accurate assessment to government and emergency management officials, giving them time to evacuate an area or take other steps to save lives.

The volcano SWAT team won acclaim in 1991 when it was sent to the Philippines to assess Mount Pinatubo, a volcano that had not erupted in several hundred years. The scientists accurately predicted that the volcano was going to explode and warned military and government officials, who in turn evacuated 35,000 Filipinos and 15,000 U.S. military personnel two days before Pinatubo violently erupted. Because of the warning, the team saved more than 5,000 lives and prevented more than a quarter-billion dollars in property damage.

More information about the Volcano Disaster Assistance Program can be found at **volcanoes.usgs.gov/About/Where/VDAP/main.html.**

set up the Cascades Volcano Observatory (CVO) in Vancouver, Washington. The staff continually monitors the Cascade volcanoes with remote sensors on the peaks as well as through direct observations and measurements. They use laser beams, electronic tilt meters, and standard surveying techniques to look for any signs of deformation that may occur when magma rises beneath a volcano. They also have been using satellite-based technologies, such as the Global Positioning System and radar, to detect even the smallest of changes.

For example, researchers with the Geological Survey used a technology called Interferometric Synthetic Aperture Radar aboard satellites to detect a swelling of the earth near the South Sister volcano in central Oregon between 1996 and 2000. The satellite images revealed that a region 9 to 12 miles across had risen about 4 inches over that four-year span. Continuing observations show that the area continues to rise about 1 inch per year.

Researchers with the CVO also have developed sensors that can detect lahars moving down a mountain, giving time for warnings to be issued to communities on or near the flanks of volcanoes. The detectors are in place in drainages on Mount Rainier, which has unleashed repeated massive lahars in past centuries over areas that people now live on.

By studying the geological events that led up to the eruption of Mount St. Helens—such as swarms of thousands of small earthquakes—scientists are better able to predict when a volcano is awakening. The Pacific Northwest Seismograph Network based at the University of Washington in Seattle also keeps track of earthquakes throughout the region, including those occurring at the volcanoes.

In the past few years, the CVO has produced detailed reports about the hazards that Hood, Rainier, and other Cascade volcanoes pose to the public. Emergency management officials throughout the Northwest are using the reports to plan what steps to take in the event a volcano does appear to be awakening.

For more information:

USGS Volcano Hazards Program
volcanoes.usgs.gov

MOUNT BAKER

The Nooksack tribe named it *quick-sman-ik,* or "white steep mountain," a befitting name for the gleaming peak that towers over the northwest Washington landscape. Now called Mount Baker, it was definitely a white steep mountain in the winter of 1998–99, when its 1,140 inches of snow set a world record for the most snowfall ever measured in a single season.

Even without the snow, the 10,778-foot peak's thirteen major glaciers make it sparkle throughout the year. Excluding the heavily glaciated and much larger Mount Rainier, Baker has more snow and ice than all of the other Cascade peaks combined.

The ice-mantled mountain, only 15 miles south of the Canadian border, is conspicuous to travelers along Interstate 5 and is an ever-present neighbor of the residents of Bellingham, Washington, which lies about 30 miles west of the peak.

Spanish sailors who explored the Strait of Juan de Fuca in 1790 were the first to document seeing the peak, calling it La Montana del Carmelo. Just two years later, other Spanish explorers observed Mount Baker erupting.

British naval explorer George Vancouver also arrived in 1792, naming the mountain after Third Lieutenant Joseph Baker. Though Vancouver's expedition didn't see the volcano erupting, Baker wrote of it in his journal, describing how "a very high conspicuous craggy mountain . . . presented itself, towering above the clouds."

Mount Baker is one of the youngest of the Cascade volcanoes. Its present cone formed only in the past 30,000 years, atop an ancient volcano

Scientists sample fumaroles at Mount Baker's summit.

called Black Buttes that was active about 400,000 years ago. Two of the craggy Black Buttes peaks can be seen just west of the mountain's summit.

Thick ice sheets eroded much of the mountain's older formations during the last ice age, which ended about 15,000 years ago. While scientists believe Baker hasn't had a major eruption in the past 14,000 years, it hasn't been totally quiet. About 6,600 years ago a flare-up near the now ice-filled crater at the peak's summit spewed ash more than 20 miles to the northeast.

Closer to our own time, Sherman Crater, about 2,500 feet south of the summit, is an active vent that geologists believe formed in the 1700s or the early 1800s. The crater's present shape can be traced to an 1843 eruption that sent two lahars, or debris flows, down the mountain's east flank. In 1891 another lahar raced 6 miles down the mountain.

Mount Baker became the center of volcanic attention for a year beginning in March 1975 when gas and steam began pouring from Sherman Crater. One large surge of steam rocketed 2,500 feet into the air, and heated material ejected from the crater onto nearby glaciers caused several small lahars. Acidic water flowed into Baker Lake, on the mountain's east side, for several months.

Scientists intensively monitored the mountain as concerns grew about a possible explosive eruption. But the peak quieted after a year, and no similar activity has been observed since.

Based on its history, the primary hazards Mount Baker presents are from lahars and avalanches, which may happen whether the volcano is erupting or not. In a 1995 report scientists said they were especially worried about the potential of a debris flow or pyroclastic flow entering Baker Lake and displacing enough water to either overtop Upper Baker Dam or cause failure of the dam. The report stated: "If Baker Dam should fail, the resulting debris flow or flood would most likely affect the entire Skagit flood plain to Puget Sound." And although there are no signs that magma is rising beneath the volcano, the report concludes that "someday in the future it will surely become restless again."

The mountain is part of the 117,500-acre Mount Baker Wilderness Area, set aside by Congress in 1984. It also is included in the Mount Baker–Snoqualmie National Forest, which extends more than 140 miles along the western slopes of the Cascades from the Canadian border to the northern boundary of Mount Rainier National Park.

Mount Baker also borders the remote North Cascades National Park, created by Congress in 1968. Within the park's 505,000 acres are numerous spectacular nonvolcanic mountains with 318 glaciers—more than half of all the glaciers in the lower forty-eight states.

North Cascades National Park is one of America's most mountainous parks. Its tallest peaks include 9,200-foot Goode Mountain, 9,127-foot Mount Shuksan, 9,087-foot Mount Logan, 9,080-foot Mount Buckner, 8,975-foot Meashchie Peak, 8,970-foot Black Peak, and 8,956-foot Mount Redoubt. The park provides a variety of activities for visitors, from leisurely hikes to challenging mountain climbing.

For more information:

Mount Baker–Snoqualmie National Forest

21905 Sixty-fourth Avenue W
Mountlake Terrace, WA 98043
(425) 775–9702, (800) 627–0062
www.fs.fed.us/r6/mbs

North Cascades National Park

2105 State Route 20
Sedro-Woolley, WA 98284
(360) 856–5700
www.nps.gov/noca

GLACIER PEAK

For those looking to get away from it all, Glacier Peak in northwest Washington is the place to go. It seems like the loneliest of the major Cascade volcanoes, even though it's only 70 miles northeast of Seattle. At a relatively small 10,541 feet, Glacier Peak isn't prominently visible from any urban area, unlike most of the range's other mountains. Because the remote mountain receives little attention, it's a perfect spot for any traveler looking for solitude. But visitors should be prepared: Unlike the other Cascade volcanoes, Glacier Peak has no roads that reach its flanks, with a hike of several miles required to reach the base.

The rugged mountain is part of the Glacier Peak Wilderness Area, which covers more than a half-million acres adjacent to North Cascades National Park. Some call it the last of North America's "wild" mountains.

Settlers didn't know it was a volcano until the 1850s, when Native Americans told naturalist George Gibbs that "another smaller peak to the north of Mount Rainier once smoked." The first white man to document seeing Glacier Peak was Daniel Linsley, who was doing survey work in 1870 for a possible route for the Northern Pacific Railroad. The mountain didn't get labeled on a published map until 1898.

The mountain's name is quite fitting, since fourteen major glaciers coat its flanks in ice, sporting a variety of colorful names such as Honeycomb, Scimitar, Ermine, Chocolate, Dusty, and Cool. Honeycomb Glacier is one of the longest (2.5 miles) and largest (1.3 square miles) glaciers in the North Cascades.

Glacier Peak has erupted more than a dozen times in the past 14,000 years.

Despite its isolation, Glacier Peak gets plenty of respect and attention as a volcano. With the exception of Mount St. Helens, the mountain has produced more and larger explosive eruptions than any other Washington volcano. In the past 14,000 years, Glacier Peak has erupted at least a dozen times, making it one of the Cascades' more active volcanoes. The most recent eruption occurred about 300 years ago. Major eruptions occurred at Glacier over the course of a few centuries around 6,000 and 13,000 years ago, generating numerous pyroclastic flows (hot, dry flows of gas, rock, and ash) and lahars (fast-moving mudflows).

The older period of activity began just after ice-age glaciers had retreated from the Pacific Northwest. During a 600-year period, the volcano generated nine ash eruptions, including one that expelled five times as much tephra—rock fragments as fine as ash—as the eruption of Mount St. Helens. Winds carried the ejected material more than 600 miles to the east, placing an inch-thick layer of tephra in western Montana. Ash from eruptions at Glacier has been found as far east as Wyoming.

During the 13,000-year-old eruptive period, the volcano propelled pyroclastic flows nearly 10 miles down its flanks and sent dozens of lahars down the White Chuck, Suiattle, and Sauk Rivers. Additional lahars sped down to the Skagit River and the Stillaguamish River all the way to Puget Sound nearly 70 miles away. The volcano also churned out lahars that went down the Skagit River to Puget Sound during the eruptive period 6,000 years ago. The debris and mud swept over what are now the towns of Sedro-Woolley, Burlington, and Mount Vernon.

About 1,800 years ago eruptions again caused lahars to reach the sea. More recently Glacier Peak has produced several smaller ash eruptions, the last one between 200 and 300 years ago. Large eruptions have produced domes of lava from Glacier Peak's vent sites. Parts of these hot domes have collapsed, producing pyroclastic flows and ash clouds. The volcano's summit is composed of the remains of ancient lava domes. Disappointment Peak, a huge "false summit" that juts up near the mountain's peak, also is made up of remnants of lava domes.

Despite its isolation, scientists say that even minor eruptions could send lahars down its flanks, inundating river valleys down to Puget Sound, an area that is attracting more people and development. The lahars could damage highways and bridges, bury houses, and block river channels. Ashfall also could disrupt air transportation and damage machinery and electrical equipment.

Glacier Peak shows no signs of awakening in the near future, however. Scientists with the Cascades Volcano Observatory estimate that the probability of a new eruption in any given year is only one in a thousand. They nevertheless warn that this volcano is worth keeping an eye on.

For more information:

Okanogan and Wenatchee National Forests
215 Melody Lane
Wenatchee, WA 98801
(509) 664–9200
www.fs.fed.us/r6/wenatchee

USGS Cascades Volcano Observatory
vulcan.wr.usgs.gov/Volcanoes/GlacierPeak

MOUNT RAINIER

First-time visitors to Seattle are often awestruck when they see majestic Mount Rainier rising 65 miles to the southeast. The towering mountain dominates the landscape, making even Seattle's impressive man-made skyline seem puny in comparison.

Mount Rainier reigns as the king of the Cascade Range. At 14,410 feet, it is more than 2,000 feet taller than Mount Adams, its nearest Washington State rival in height, and about 250 feet taller than Mount Shasta, the second highest Cascade volcano. Rainier's enormity and beauty led Congress to make the mountain the centerpiece of a national park in 1899. Today that park encompasses 368 square miles, 97 percent of it designated as wilderness.

Long before European explorers came to the area, Native Americans lived for thousands of years in the mountain's shadows. They called it a variety of names, including *tahoma, takhoma,* and *ta-co-bet*—meaning "snowy peak," "big mountain," and "place where the waters begin." British naval explorer George Vancouver, who sailed into Puget Sound in 1792, named it for his friend Rear Admiral Peter Rainier.

Mount Rainier isn't just a pretty face for postcards and camera-toting tourists, however. The majestic mountain is considered the most dangerous Cascade volcano, with its seemingly tranquil wildflower- and snow-covered slopes disguising the peak's perils.

The active but resting volcano last erupted 150 years ago, but the mountain poses a growing threat to people and property as urban development spreads into the five river valleys that lead away from the mountain. Rainier

Mount Rainier has twenty-six glaciers on its steep slopes.

begins to rise only about 25 miles from the Seattle-Tacoma metropolitan area and its population of two-and-a-half million people. Part of the volcano's watershed empties into Puget Sound and into the Columbia River.

Scientists have warned for years that the major danger posed by Rainier is from huge debris flows, which have swept down these valleys many times in the past 10,000 years. And, scarily, it doesn't necessarily take an eruption to set them off. Recent reports by the U.S. Geological Survey not only reconfirm the danger, but say a house built downstream in any of these areas is twenty-seven times more likely to be destroyed or damaged by one of these debris flows than it is to be destroyed or damaged by fire. These debris flows, also called lahars, have been compared to liquid concrete. Lahars typically move between 25 and 50 mph down a mountainside, giving little time to alert people to get to safety.

Lahars are a major threat to people living near Rainier, especially when you consider that the volcano's twenty-six glaciers contain more than five times as much snow and ice as all of the other Cascade volcanoes combined. Even only a small part of the ice melted by ash or lava would produce enough water to trigger enormous lahars.

It's important to remember that a major debris flow can form without a volcanic eruption. Over time the volcano's once-hard interior rock is transformed into soft clay by centuries of exposure to hot acidic gases. An earthquake or an oversaturation of water from intense warm rains or from a high rate of glacial melting in hot weather could collapse this structurally weakened rock.

Rainier has experienced large landslide-initiated flows every 500 to 1,000 years, with the last major flow of this type occurring about 500 years ago. Past lahars, up to a dozen over the last 6,800 years, filled the mountain's previous valleys with extensive debris-flow deposits—land on which more than 100,000 people now live, with the numbers growing each year.

The lahar 500 years ago—known as the Electron Mudflow—broke loose from the upper west side of Rainier and traveled down the Puyallup River valley. It covered the site of the present-day town of Orting northwest of the mountain with a flow of mud and debris 2 miles wide and 20 to 30 feet thick.

Scientists say a future flow could reach all the way to Puget Sound. The largest known lahar from Mount Rainier, known as the Osceola Mudflow, traveled down the White River drainage system for nearly 70 miles about 5,700 years ago. The flow reached as far as the Seattle suburb of Kent and to Puget Sound at Tacoma.

Smaller flows are common on Rainier. In the past fifty years, about two dozen debris flows and glacial outburst floods have occurred here, traveling as far as 10 miles from their origin. Glacial outburst floods, the sudden release of water from glaciers, are turned into dangerous debris flows as they pick up mud and rocks. These floods are more likely to happen during the very hot weather of late summer or early fall. In 1947 a flood and resulting debris flow traveled from Kautz Glacier on the volcano's southern flank 5.5 miles down Kautz Creek, burying the Nisqually-Longmire Road under 28 feet of mud and rock. Dozens of smaller debris flows have occurred in recent years, especially on South Tahoma Glacier, where a series of flows between 1967 and the early 1990s forced the closure of the park's western road. Debris flows also were triggered in August 2001 by melting water pouring from Kautz Glacier, with car-size boulders sent hurtling down Van Trump Creek.

Glaciers cover more than 34 square miles of Mount Rainier. Of all the glaciers in the contiguous United States, the peak's Emmons Glacier has the largest surface area, with 4.3 square miles. Carbon Glacier is the longest at 5.7 miles, the thickest at 700 feet, and has the most volume at 0.2 cubic mile.

Scientists warn that a warming climate has caused Mount Rainier's massive glaciers, along with those on other Cascade peaks, to wane in recent decades. The present-day Rainier probably began forming about 700,000

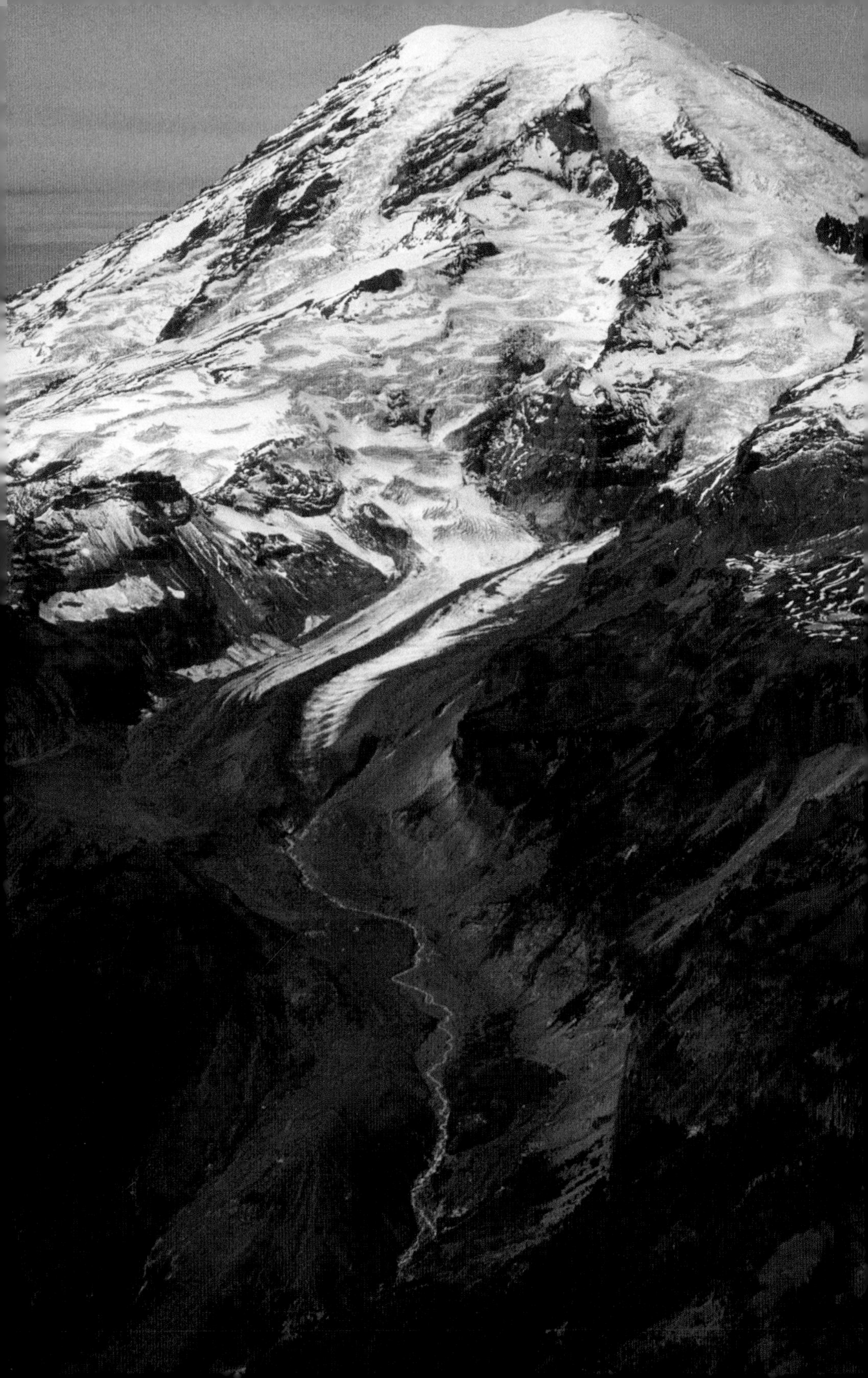

years ago. By the end of the last ice age about 10,000 years ago, eruptions eventually constructed the volcano's cone that now stands about 7,000 feet above its surroundings.

Mount Rainier may have reached nearly 16,000 feet high at one point, but a violent event 5,700 years ago sliced nearly 2,000 feet off its summit. A huge landslide—perhaps triggered by magma forcing its way into the volcano—swept down the mountain's northeast side, creating a deep horseshoe-shaped crater. The collapse generated the Osceola Mudflow. The last significant major eruption, about 2,200 to 2,500 years ago, generated a lava cone at the mountain's summit that fills much of the crater.

Although there are a few dozen earthquakes reported in the vicinity of Mount Rainier each year, the volcano does not show any signs of erupting soon. But because of the multiple volcanic hazards that it poses to a large population, it is considered America's most dangerous volcano, one geologists say needs to be continually, carefully watched.

For more information:

Mount Rainier National Park
Tahoma Woods, Star Route
Ashford, WA 98304-9751
(360) 569–2211, ext. 3314
www.nps.gov/mora

USGS Cascades Volcano Observatory
vulcan.wr.usgs.gov/Volcanoes/Rainier

◄ Floods from Mount Rainier's glaciers pose a threat to those downstream.

MOUNT ST. HELENS

"Vancouver, Vancouver. This is it! Is the transmitter working?"

David A. Johnston frantically radioed his final words on May 18, 1980, alerting colleagues that Mount St. Helens was awakening in a fury. From a monitoring post 5.5 miles north of the summit, Johnston watched as the youngest and most active volcano in the Cascades provided a twenty-four-megaton reminder of nature's power. Within moments the thirty-year-old scientist with the U.S. Geological Survey became one of fifty-seven people to die in the eruption.

The cataclysmic chain of events began at about 8:32 A.M. that tragic Sunday. Loosened by a 5.1-magnitude quake about a mile under the volcano, the upper north flank—weakened and deformed by two months of eruptive activity—collapsed in the largest avalanche in recorded history. Nearly three-fourths of a cubic mile of rock, mud, ice, snow, and water surged northward across the Spirit Lake basin, over the ridge where Johnston was perched, and then westward 17 miles down the North Fork of the Toutle River.

The slide uncorked the volcano's internal pressure, unleashing a lateral blast of gas and ash—as hot as 868 degrees and with a top speed of 670 mph—that toppled trees as far as 13 miles away. An enormous column of ash rose 13 miles into the atmosphere and lasted nine hours. About 540 million tons of ash spread over 22,000 square miles in three days and circled the earth in fifteen days.

Debris flows the consistency of wet concrete raced down almost all of the valleys draining the volcano. The largest flow started later in the day and

Mount St. Helens awakened with steam explosions and earthquakes in March 1980.

went down the North Fork of the Toutle River. The flow—more than 30 feet deep and 500 yards wide—reached a speed of 27 mph and wound up disrupting shipping traffic on the Columbia River for three months.

No one envisioned the powerful forces that would knock out Mount St. Helens's north side that Sunday. Until that day most people viewed the rugged Cascades as landscape-beautifying peaks rather than landscape-altering volcanoes. But the catastrophic eruption blew away that peaceful notion, slicing 1,313 feet of its summit, which now stands at 8,364 feet above sea level at its highest point.

Beginning March 16, 1980, the mountain began to awaken from a 123-year sleep. Thousands of earthquakes occurred during the next two months as magma slowly worked its way up into the volcano, with the molten rock fracturing the older volcanic cone. The magma heated the inside of the mountain, and on March 27 an eruption of steam and ash blew a 250-foot-wide crater in the summit. Almost immediately a bulge began to develop on the northern slope, and its walls continued to swell and steepen, making a landslide inevitable. The 5.1-magnitude earthquake on the morning of May

THE AVID **SCIENTIST**

Everyone thought David A. Johnston was safe at his observation post near the crest of the first ridge north of the Toutle River. But no one envisioned the powerful forces that would devastate Mount St. Helens's north side on Sunday, May 18, 1980.

Johnston, a thirty-year-old scientist in charge of volcanic-gas studies at Mount St. Helens, was the only scientist to die in the eruption that took fifty-six other lives. Dr. Stephen D. Malone, research professor of geophysics at the University of Washington, describes Johnston as a "fire dog." "When something interesting was going on, he wanted to be there, the first involved. He was very good at his work."

Johnston was working at the U.S. Geological Survey in Menlo Park, California, when the mountain began awakening in the spring of 1980. He happened to be in Seattle at the University of Washington, where he had received his doctorate, when the volcano's first flurry of earthquakes were recorded on March 16 by the university's seismology lab. "He heard about them almost immediately, and he came in to see me," Malone says, "so I put him to work." When news reporters asked for a volcano expert to escort them to the mountain, Johnston headed out with them and joined other scientists arriving to monitor it.

A native of Illinois, Johnston received his degree in geology in 1971 from the University of Illinois at Urbana. After completing his doctorate at the University of Washington in 1978, he joined the Geological Survey, where he was assigned to expand the program for monitoring volcanic emissions in Alaska and the Cascades.

Richard B. Waitt, a geologist at the Cascades Volcano Observatory and one of the first scientists sent to monitor the volcano, said the young volcanologist "was a real spark plug. . . . We all loved Dave. He was just a bundle of energy and excitement and knowledge."

On the ridge where he died, an observatory now bears Johnston's name. It's the closest visitor site to the crater.

Mount St. Helens exploded in fury on May 18, 1980.

18 sent the unstable north flank plunging away, with the slide crashing into Spirit Lake and sending a wall of water as high as 800 feet up the surrounding ridges. The eruption damaged 27 bridges and nearly 200 homes and killed enough trees to build 300,000 small houses.

Smaller eruptions continued to occur at the volcano in the following years. The last dome-building event, in which magma reached the surface and added to the pile of lava on the crater floor, was in October 1986. The 3,480-foot-wide dome stands about 826 feet above the crater floor. The last time the volcano spewed a plume of steam and ash from the dome was in February 1991.

Though even volcano scientists were stunned by the ferocity of the 1980 event, it came as no surprise to the geologists who had studied the mountain. Dwight "Rocky" Crandell and Donald Mullineaux of the Geological Survey had warned in a 1978 report that Mount St. Helens posed a considerable threat.

Mount St. Helens is the youngster of the Cascade volcanoes, throwing explosive tantrums since its birth about 40,000 years ago. The mountain's growth spurt began about 13,000 years ago at the end of the last ice age, with at least eight significant eruptive periods before the 1980 blast. The length of these intermittent episodes varied, with the shortest lasting about 100 years and the longest 5,000 years.

The volcano's restless nature was well known to the region's earliest inhabitants. Some tribes, such as the Salish and Klickitat, called it *loo-wit*

THE
GLACIER

The new glacier that is growing around Mount St. Helens's lava dome is providing scientists with an exciting opportunity to watch nature in action.

A glacier is a large body of ice that survives from year to year and shows evidence of movement. Scientists with the U.S. Geological Survey are calling the one at Mount St. Helens an "incipient glacier." As Jon J. Major, a hydrologist at the USGS's Cascades Volcano Observatory, explains, "The lower parts of it are true ice, and even on the front of the body on the west side of the crater you see true ice. Then there are several places where crevasses are present, and those are indicative of motion." The mountain's south wall keeps the area in shade for all but a couple of hours a day during the summer, and it receives virtually no sunshine in the winter.

The location of the new glacier in the Mount St. Helens crater

Amateur cave explorers with the International Glaciospeleological Survey have been investigating the ice caves at the bottom of the structure since 1982. According to the group, the glacier ice isn't apparent from a distance because it is hidden by snow and rock. They say snow stacking higher each year has compressed the lower layers (visible in the caves) into dense crystalline ice. Another indication that a new glacier is forming are the crevasses and flow textures in small areas of ice visible on the south crater wall.

Steve P. Schilling, a hydrologist at the Cascades Volcano Observatory, says the rock off the crater walls also is helping to protect the snow and ice from melting. Schilling is using a computer technology called digital photogrammetry to investigate the massive slab. He scans aerial photographs taken in the past twenty years into the computer so he can compare digital elevation models. "You can get an estimate of how much erosion has taken place on the crater walls, how much the dome has grown, and how much snow, ice, and rock has accumulated in the moat of the crater," Schilling says. "And if you get successive years of photographs, you can get at some aspects of how fast these things happened."

The information will be used to estimate the danger posed by the accumulation of snow and ice if an eruption occurs in the crater. In 1982 a small eruption rapidly melted the crater's snowpack, resulting in a mudflow that swept down the Toutle River and into the Cowlitz River. The U.S. Army Corps of Engineers built a sediment-retention dam in 1987 on the North Fork of the Toutle, to lessen the threat to the people downstream.

lat-kla or *louwala-clough,* meaning "fire mountain" or "smoking mountain." Tree-ring dating shows that beginning in 1480, the volcano began a series of explosive eruptions that lasted for nearly a century.

On May 19, 1792, Captain George Vancouver of the British navy became the first European to record observing the peak, though he never saw it in action. He named it after Alleyne Fitzherbert, Britain's ambassador to Spain, who held the title Baron St. Helens.

The mountain erupted soon thereafter, in 1800, five years before the arrival of the Lewis and Clark expedition. Beginning in 1831 the volcano occasionally popped off until 1857, when it became docile for the following 123 years.

Earthquakes continue to rumble under the mountain, as pressurized pockets of gas from the magma beneath the crater cause the surrounding rock to break. Flurries of quakes were detected for brief periods in 1995 and 1998. Another swarm of earthquakes in November 2001 was likely related to

A scientist looks at trees hit by the lahars of May 18, 1980. ➤

THE RETURN OF LIFE

Nearly a quarter-century after the 1980 blast blew away, buried, or burned plant and animal life on Mount St. Helens, a biological eruption is occurring on the volcano's landscape. Signs of nature gradually repairing itself are apparent everywhere.

Bear tracks in the volcanic ash

On the stark pumice plain that stretches 4 miles northward from the crater, prairie lupines and burrowing northern pocket gophers are preparing the soil for newcomers. Tall willows and small trees have emerged in patches, attracting birds and insects along with mammals that deposit new plant seeds.

Large animals such as Roosevelt elk and Columbia black-tailed deer have returned to the blast zone, leaving grass and other plant seeds in their scat. Smaller mammals and birds such as Canada geese also bring in seeds.

Ecologists surveying small plants and animals have found that many species populations have become fixed communities on the windblown plain, helping to set the stage for explosive growth. Thousands of young conifers also are emerging and will eventually change the face of the mountain. A small willow grove has emerged along a spring in one area of the pumice plain, attracting a dozen small-mammal species such as deer mice and golden-mantled ground squirrels, a half-dozen bird species, and four amphibian species.

A plant that's key to the comeback of life is the prairie lupine, which has been called "the star" of Mount St. Helens. Just two years after the eruption, scientists spotted lupines pushing up through the powdery gray pumice. Thousands of the blue-flowered plants now cover large areas, and with the nitrogen and carbon they add to the nutrient-poor soil, they're helping to open the way for dozens of other types of plants.

Even though the lupine has had a dramatic impact on the still-barren-looking landscape, the flowers weren't spreading as rapidly as scientists had anticipated. Biologist John Bishop, a professor at Washington State University in Vancouver, wanted to know why. After studying the lupine's role on Mount St. Helens for several years, he's concluded that lupine-loving butterfly and moth caterpillars are to blame. Bishop says the insects are devastating to the plants, with one caterpillar species eating the plant's seeds while another bores into its stems. A third caterpillar weaves all the leaves together and turns them yellow.

The expansion of vegetation has been more a pattern of fits and starts than a steady, predictable cycle of growth. Droughts and floods, extremes in cold and heat, voracious insects, and foraging animals can take a toll. Scientists point out that time and chance continue to play an important role in the future of plant and animal life on Mount St. Helens.

increased rainfall. As water percolates into the lava dome and crater floor, fractures slip and small landslides and debris flows occur, sometimes with small steam explosions hurling rocks a few hundred yards.

The 1980 eruption devastated the volcano's glaciers, obliterating 70 percent of the peak's ice. The sprawling glaciers Loowit and Leschi vanished, nearly all of Wishbone Glacier disappeared, and large parts of seven other glaciers were also destroyed. But now a new glacier has formed within the crater around the lava dome, protected by the high, steep crater walls south of the dome. The horseshoe-shaped pile of snow, ice, and rockfall has been building the past two decades, giving scientists a chance to observe the glacier's formation. The developing glacier is an anomaly in the Cascades, where glaciers have been retreating for decades.

The massive body of ice, snow, and rock on the crater floor is more than 650 feet thick and contains roughly 157 million cubic yards of material, enough to fill the Empire State Building nearly 115 times. Rock debris shed off the crater walls is thought to make up about one-third of the structure, the remainder being snow and ice. The volume of snow and ice that's locked up in that body is 30 to 40 percent of the volume of glacial ice that was on the mountain prior to the eruption.

Mount St. Helens has become a vast, living science laboratory, with a 110,000-acre national monument set aside in 1982 to preserve, study, and explain the eruption and its aftermath. Only 50 miles northeast of the Portland metropolitan area, it is also a major tourist attraction, with more than a half-million people visiting the mountain during the spring and summer.

The USDA Forest Service, which oversees the national monument, has two visitor centers along Spirit Lake Highway, which winds up the Toutle River Valley from Interstate 5. A third visitor center is operated by Washington State Parks. The monument is part of the 1.3-million-acre Gifford Pinchot National Forest. Hiking is a popular activity in the national monument, along with climbing the volcano's south side.

Though scientists continue to monitor the volcano for any signs of activity, they say a landslide and blast similar to the 1980 event isn't likely. Most agree that floods and debris flows pose the greatest risks from future volcanic activity. Mount St. Helens has stayed relatively quiet in recent years, but given its restless history, the fidgety youngster of the Cascades undoubtedly will erupt again.

For more information:

Mount St. Helens National Volcanic Monument
Monument Headquarters
42218 Northeast Yale Bridge Road
Amboy, WA 98601
(360) 449–7800
www.fs.fed.us/gpnf/mshnvm

USGS Cascades Volcano Observatory
vulcan.wr.usgs.gov/Volcanoes/MSH

MOUNT ADAMS

Mount Adams has a tough time garnering much recognition. It's been one of the more quiet Cascade volcanoes in the past few millennia, and its remote setting draws relatively few visitors. Even Lewis and Clark mistakenly thought the huge mountain was Mount St. Helens as they paddled down the Columbia River, 35 miles to the south, in 1805.

Still, Mount Adams is a majestic volcano that towers over south-central Washington, a formidable peak to rival other Cascade mountains. With its summit at 12,276 feet above sea level, Mount Adams is the third highest peak in the Cascades and the second highest peak in Washington behind Mount Rainier.

Despite its prominence, Mount Adams receives scant attention when compared to more active Mount St. Helens 35 miles to the west or higher Mount Rainier 50 miles to the northwest. The volcano's isolation and lack of recent eruptions make it a largely ignored behemoth. The upper slopes of Mount Adams are among the most rugged and least visited parts of the Cascade Range, making it a destination for many serious climbers and hikers. Part of the Gifford Pinchot National Forest, the mountain is protected within the 46,000-acre Mount Adams Wilderness Area and also within The Yakama Indian Reservation.

Mount Adams is described in some Native American legends as a grieving loner, Pahto. One tale describes how Pahto and his brother Wy'east (Mount Hood) waged a fierce battle for the affections of the beautiful maiden La-wa-la-clough (Mount St. Helens). In anger, the Great Spirit, Sahale, struck down the three competing lovers but raised a mountain peak where each

Mount Adams towers over the surrounding countryside.

fell. Because of her beauty, La-wa-la-clough became a perfect symmetrical cone (until the explosive 1980 decapitation). Wy'east lifts his head in pride, but Pahto hangs his head forever in sorrow. The description is befitting of the volcano, as its profile is rather stubby when compared to Mount Hood's majestic pinnacle 60 miles to the south.

Even its current name is the result of an error. In the mid-nineteenth century an unsuccessful plan to name the Cascade volcanoes in honor of presidents led to a mapmaker accidentally placing the name Adams (for President John Adams) on the peak. The presidential name actually had been intended for Mount Hood. To add insult to injury, the mountain now known as Adams had been excluded from the original naming scheme.

But Mount Adams definitely deserves a place among the Cascades' most significant peaks. In volume, Adams ranks only behind Mount Shasta among Cascade volcanoes. One estimate indicates that its volcanic material would fill a hole 15 miles long and 10 miles wide with nearly 1,000 feet of debris. The glistening volcano also has a dozen glaciers adorning its upper slopes.

Volcanic activity in the area began nearly a million years ago, with Adams beginning to take shape about 460,000 years ago. Most of its main cone is younger than 220,000 years, and much of the mountain above the timberline was rapidly built by eruptions between 12,000 and 30,000 years ago.

Snow-wrapped Mount Adams shows its beauty at sunset.

At least eight large lava flows have spewed from vents on the mountain in the past 10,000 years, covering an area of about 20 square miles. The most recent lava flow occurred between 3,500 and 6,800 years ago. A series of small eruptions occurred about 1,000 years ago, but the volcano has been quiet since then.

Despite appearing peaceful, the slopes of Mount Adams can be hazardous, especially from avalanches. In 1921 about 5 million cubic yards of rock plummeted 4 miles down the Salt Creek valley on the peak's southwest flank. The avalanche, which originated at the appropriately named Avalanche Glacier, also had enough water to produce small lahars, fast-moving mudflows. Similar ice-and-rock avalanches ripped down the mountain in August and October of 1997. The summer avalanche also originated at Avalanche Glacier, with an estimated 6.5 million cubic yards of debris moving 3 miles down the mountain. The October avalanche—similar in size to the August slide—ripped 3 miles down the east side of the mountain on October 20, 1997. Wet weather, not earthquake or volcanic activity, caused the avalanches, a clear demonstration that steep Cascade mountains always pose a danger whether they're active or not. Scientists point out that landslides can attain speeds that exceed 100 mph, and the largest avalanches can sweep down valleys more than 30 miles before stopping, destroying everything in their paths and choking river valleys with sediment.

The 1997 avalanches pale in comparison to ancient slides that have occurred on Mount Adams. About 6,000 years ago a huge avalanche with 90 million cubic yards of debris hurtled down its slopes. It produced speeding lahars that inundated the Trout Lake lowland to the south and continued down the White Salmon River valley more than 35 miles south of the mountain. The lahar deposit in the lowland measures up to 65 feet thick in some places. The affected area is now the site of several communities—a reminder that avalanches and lahars on Mount Adams pose the most serious hazards to people and property near the volcano.

Because Mount Adams has been one of the least active Cascade volcanoes in the past few thousand years, scientists don't foresee the mountain erupting in the near future. But the volcano isn't dead, they warn, and will likely stir to life again at some point in the future.

For more information:

Gifford Pinchot National Forest
10600 Northeast Fifty-first Circle
Vancouver, WA 98682
(360) 891–5000
www.fs.fed.us/gpnf/recreation/mount-adams

Mount Adams Ranger District
2455 Highway 141
Trout Lake, WA 98650
(509) 395–3400

USGS Cascades Volcano Observatory
vulcan.wr.usgs.gov/Volcanoes/Adams

MOUNT HOOD

To early Native Americans who once feared and revered this craggy volcano, Oregon's loftiest mountain represented a powerful warrior god they called Wy'east. Now bearing the name of a British naval officer, Mount Hood continues to awe those who view the spectacular landmark.

At 11,240 feet, Mount Hood is the fourth highest peak in the Cascades. The snowcapped edifice dominates the landscape of northwest Oregon, providing a picture-postcard backdrop to the city of Portland only 60 miles to the west.

The easily accessible volcano is one of the world's most climbed mountains and draws tens of thousands of skiers, backpackers, and tourists to its slopes each year. Timberline, a picturesque ski lodge, was built on its southern flank by the Works Progress Administration in the 1930s and is still open to visitors.

Learning more about the mountain's restless past provides clues about what the volcano might do when it reawakens, which scientists say is sure to happen. Mount Hood is between 500,000 and 700,000 years old and sits atop a smaller, older volcano. After an active period of eruptions during the last ice age, between 15,000 and 30,000 years ago, Oregon's tallest peak quieted until a major eruptive period about 1,500 years ago. The volcano then calmed down for more than a millennium before awakening again about 200 years ago.

The first humans to gaze on the mountain some 12,000 years ago would have seen a solitary peak soaring above the countryside. Native American legends recount how Wy'east and his brother Pahto (Mount Adams) clashed for the maiden La-wa-la-clough (Mount St. Helens). The tales speak of how

Mount Hood towers over Lost Lake.

Wy'east hurled fiery rocks and spewed streams of fire, making it clear that the region's early inhabitants witnessed its volcanic activity.

Later arrivals also were captivated by the majestic sight. The first recorded description of the volcano came in 1792 from Lieutenant William Robert Broughton, a British officer with Captain George Vancouver's exploratory voyage to the Pacific Northwest. "A very distant high snowy mountain now appeared rising beautifully conspicuous in the midst of an extensive tract of low, or moderately elevated land," Broughton wrote, naming the mountain for Admiral Samuel Hood.

Thirteen years later Meriwether Lewis and William Clark also took note of the peak during their historic expedition to the Pacific Ocean down the Columbia River a few miles north of the mountain. Although the volcano was relatively quiet when they visited, Lewis and Clark's descriptions of what is now the Sandy River are similar to what was seen in the sediment-laden Toutle River after the 1980 eruption of Mount St. Helens. While neither the British nor the American explorers observed any restlessness from Mount Hood, their visits occurred during the peak's last significant episode of activity, called the "Old Maid" eruptive period, which lasted from about 1780 to at least 1810.

Scientists recently pinpointed the date of a key event of that period, when a huge lahar—a hot, concretelike soup of mud and debris—swept down the mountain's Sandy and Zigzag Rivers to the south and west, burying trees and the stream channel. Using the annual growth rings from trees buried or damaged by the lahar, researchers calculate that the huge mudflow occurred between the fall of 1781 and the spring of 1782.

The activity two centuries ago was centered at Crater Rock, about 700 feet below the summit on the mountain's south side. The hot, high-density flows of gas, rock, and ash known as pyroclastic flows joined with the lahars primarily on the mountain's south and west sides. Crater Rock is the remnant of the lava dome that rose from the vent site during the last eruptive period. It's similar to the 826-foot-high lava dome in Mount St. Helens's crater, except Hood's lava emerged in a steeply sloping crater rather than onto a flat floor.

Because the dome grew on a steep slope, when some extremely hot chunks broke off, they triggered a pyroclastic flow at thousands of degrees Fahrenheit. This high-density mixture of rock, ash, and gas in turn triggered a destructive lahar as it quickly melted snow and ice on the mountain. The fast-moving lahar rushed down the mountainside, burying forests in up to 30 feet of debris along the Zigzag and Sandy Rivers for at least 22 miles.

Lahars present the biggest risk if Mount Hood stirs again. Scientists with the U.S. Geological Survey's Cascades Volcano Observatory warn that future eruptions could disrupt air, river, and highway transportation; some municipal water supplies; and hydroelectric power in northwest Oregon and southwest Washington.

But when will it happen? And where on the mountain? No one can accurately determine when and where the next eruption will occur. But because the last two major eruption periods were from the Crater Rock vent on mountain's south side, geologists think that would be the most probable location for a future eruptive event. An eruption from that area would affect the more populated and developed areas along rivers to the south and west.

The next time the hot magma moves up into the volcano, however, debris avalanches could also occur on the steep north and east sides of the mountain, which might cause a new vent to open. If a vent opened on the east or north side of the summit, lahars could flow down the forks of the Hood River all the way to the Columbia River.

Although the last eruptive period ended in the early 1800s, eyewitnesses reported minor volcanic events in 1859 and 1865. In 1907 a Geological Survey topographer also reported dense steam spewing from the Crater Rock area, along with a bright glow at night.

It's important to realize that, even without eruptive activity, small lahars can be unleashed by intense rainstorms or outbursts of glacial water. Mount

Hood has a dozen glaciers and carries a heavy snowpack for much of the year, providing ample water for the sliding lahars that occur every few years down its slopes.

While scientists say there is no evidence of imminent volcanic activity, the volcano may only be sleeping. Fumarole fields, called Devil's Kitchen, near Crater Rock emit a foul rotten egg smell, indicating that gases are still churning in hot areas far below the mountain. Temperature readings on the heated ground at Devil's Kitchen at 10,500 feet reach about 190 to 200 degrees, nearly the boiling point of water at that elevation. The fumaroles serve as a reminder that Mount Hood is far from dead.

Mount Hood is unique among Cascade volcanoes because swarms of small earthquakes occasionally occur about 3.5 miles beneath its south slope. As recently as the summer of 2002, more than 250 earthquakes were detected in a three-month period, including a magnitude 4.5 quake that was the largest at Mount Hood in decades.

Scientists with the Cascades Volcano Observatory aren't certain whether the earthquakes are related to a complex system of faults or are a result of changes in the volcano's magma plumbing system. Regardless, they are proof that geologic processes beneath the mountain are dynamic and warrant continual monitoring.

For more information:

Mount Hood Information Center
65000 East Highway 26
Welches, OR 97067
(503) 622–4822, (888) 622–4822
www.mthood.info

Mount Hood National Forest
16400 Champion Way
Sandy, OR 9755
(503) 688–1700
www.fs.fed.us/r6/mthood

USGS Cascades Volcano Observatory
vulcan.wr.usgs.gov/Volcanoes/Hood

MOUNT JEFFERSON

Once the skies cleared, the band of explorers led by Meriwether Lewis and William Clark couldn't miss the snow-covered mountain 75 miles south of their Columbia River campsite. The now-famous Corps of Discovery spotted the rugged peak on March 31, 1806, as they were heading back to St. Louis after a miserable wet winter near the Pacific shoreline.

One of the exploration party's members noted the event: "We saw a high mountain laying a great distance off to the Southward of us, which appeared to be covered with snow. Our Officers named this Mountain Jefferson Mountain."

Lewis and Clark, who had been on their long journey for nearly two years, named the mountain to honor President Thomas Jefferson, who had commissioned their expedition. Lewis described the mountain, the only Cascade peak they named, as "noble." Modern explorers who gaze upon the 10,495-foot volcano would agree with that description. Mount Jefferson is Oregon's second largest mountain, only about 750 feet shorter than Mount Hood 50 miles to the north. The summit at one time may have been as high as 12,000 feet, but glaciers have slowly whittled it down to its current height.

Like Mount Hood, glacier-covered Jefferson is a classic stratovolcano, with steep sides and a sharp summit created by tens of thousands of years of eruptions. Five major glaciers—Milk Creek, Jefferson Park, Russell, Whitewater, and Waldo—adorn its steep flanks in white sheets.

The mountain can be seen by passersby from major highways in the area. Its height above the surrounding terrain makes it conspicuous, but the volcano is not easy to reach. Mount Jefferson sits both within a 107,000-acre

Mount Jefferson can be seen throughout central Oregon.

wilderness area that is managed by the USDA Forest Service and the Warm Springs Indian Reservation. Climbers say the extremely steep, rocky peak is one of the most challenging of the Cascade Mountains.

Although the rugged edifice is an imposing feature on the landscape, the volcano is dull when it comes to eruptive activity. Of the thirteen major volcanic centers in the Cascades, Mount Jefferson and Crater Lake are the only ones not to have had any volcanic action within the last 4,000 years.

The last large explosive eruption occurred between 35,000 and 100,000 years ago, when vast glaciers covered the region. Some scientists consider the mountain as volcanically dead; however, a 1999 hazards report issued by the U.S. Geological Survey stated that experiences at other volcanoes "suggests that Mount Jefferson cannot be regarded as extinct."

Even though the volcano is in a remote area, the volcanologists warn that a future eruption might produce hot flows of gas and debris that could affect river valleys and say ash plumes could pose problems to communities hundreds of miles away. Landslides also could occur on the steep slopes and transform into fast-moving lahars, cementlike flows of mud and debris. The scientists caution that lahars might well enter large reservoirs—Detroit Lake to the west and Lake Billy Chinook to the east—and give rise to waves that could cause dams to fail, allowing walls of water to inundate communities downstream.

While Mount Jefferson has been quiet for several millennia, smaller nearby volcanoes have been active in more recent times. A small volcano a few miles south-southeast of the mountain, Forked Butte, generated lava flows in 4,500 B.C. Forked Butte is just one of hundreds of smaller volcanoes that appear throughout central Oregon. Among the more notable volcanic features within a 30-mile radius south of Jefferson are Three Fingered Jack, Mount Washington, and Belknap Crater.

Three Fingered Jack is a deeply glaciated volcano about 15 miles south of Jefferson that rises to 7,841 feet above sea level. Ten miles farther south is Mount Washington's pointy pinnacle, which soars to 7,796 feet. Another 5 miles to the south is Belknap Crater at 6,874 feet. This young volcano erupted as recently as 1,500 years ago, with lava flowing 12 miles to the west. Belknap is a classic example of a shield volcano, with fluid lava flows building a gently sloping cone.

A good place to view these and other volcanic features of the central Oregon Cascades during the summer and fall is at the Dee Wright Observatory, which is in the middle of a massive lava flow. The rock structure—built by the Civilian Conservation Corps in 1935—is at the summit of McKenzie Pass on Oregon Highway 242 about 15 miles west of the town of Sisters. A paved half-mile interpretive trail winds through the lava flows, with signs describing the area's geology and history. Central Oregon's Cascades draw tens of thousands of campers, hikers, climbers, white-water rafters, and anglers every summer. Large resorts, such as Black Butte Ranch and Sunriver, offer tennis, golf, swimming, biking, and other activities.

Two centuries after Lewis and Clark first spotted Mount Jefferson off in the distance, the volcano now serves as a key centerpiece to one of the nation's most unique landscapes.

For more information:

Willamette National Forest

Detroit Ranger District
HC 73, P.O. Box 320
Mill City, OR 97360
(503) 854–3366
www.fs.fed.us/r6/willamette

USGS Cascades Volcano Observatory

vulcan.wr.usgs.gov/Volcanoes/Jefferson

THE THREE SISTERS

The satellite images tell the tale: The ground is swelling in the vicinity of the Three Sisters—North Sister, Middle Sister, and South Sister—a distinctive set of volcanoes in the central Oregon Cascades. The bulging may be slight, but it is the most striking geological change in the Cascade Range since the 1980 eruption of Mount St. Helens.

Scientists with the U.S. Geological Survey revealed the surprising news in 2001 when they found that the ground had been swelling about 1 inch each year since 1996 over an area 9 to 12 miles across just west of South Sister. Detected only with a radar instrument aboard a satellite and confirmed with the Global Positioning System and other measurements, the swelling can't be seen by the naked eye. Still, the bulging indicates there's an infusion of magma in a chamber about 3 or 4 miles beneath the surface.

The news that there was volcanic activity in the area was startling, as few earthquakes or other signs of unrest had previously been detected there. Although remotely located in a 285,000-acre wilderness area, the volcanic center still poses a potential hazard to the region. Eventually the magma could rise to the surface in an eruption, which is why scientists are keeping a very close eye on the Three Sisters.

The mountains provide a picture-postcard backdrop for a fast-growing resort area and the city of Bend about 25 miles to the east. Providing a spectacular scene as they rise above Oregon's high desert, the Three Sisters are an unusual sight for the Cascades. While all the other major Cascade volcanoes rise in relative isolation on average about 40 miles from each other,

South Sister's beauty is reflected in Sparks Lake.

these three mountains are clustered together. All three are almost the same height, with South Sister the tallest at 10,358 feet, followed by North Sister at 10,085 feet and Middle Sister at 10,047 feet.

No one is certain how the volcanoes became to be called the "Three Sisters," a name that first appeared on maps in the 1850s. Early settlers with a religious bent dubbed them Faith, Hope, and Charity (north to south), but the source of their present names remains a mystery.

South Sister is the youngest of the three and is the only one to have been volcanically active in the past four millennia. Its last two eruptions spewed lava and ash about 2,000 years ago and produced small pyroclastic flows—fast-moving flows of extremely hot gas, rock, and ash. The blasts buried an area within a mile of the vent under 7 feet of ash. Larger eruptions occurred about 15,000 to 30,000 years ago.

North Sister is the oldest of the Three Sisters, and there are no signs that the rugged volcano has erupted in more than 100,000 years. Middle Sister appears to have been active about 20,000 years ago.

Despite the lack of eruptions, a hazard report by the USGS's Cascades Volcano Observatory in 1999 warned that pyroclastic flows on South Sister and Middle Sister could melt ice and snow, resulting in fast-moving lahars—concretelike flows of mud and rocks. Because of their steep sides, the three

SATELLITE MONITORING

Instruments aboard high-flying satellites are giving scientists a new look at volcanoes. Not only are they beginning to play a key role in monitoring volcanic activity, they are also detecting subtle changes in volcanoes. A new technique for measuring tiny ground movements from orbiting satellites provided the stunning information that an area about 3 miles west of South Sister was inflating at a rate of about 1 inch per year as magma was rising far below the surface.

The scientists were able to detect the subtle uplift, which is not visible from the ground, using Interferometric Synthetic Aperture Radar, or INSAR, aboard satellites. INSAR images are taken when a satellite-borne instrument beams radar waves to the ground and receives the waves when they bounce back to the spacecraft. The instrument records the strength and time delay of the returning signal to produce images of the ground. By comparing images of the same 60-by-60-mile area at different times, they can spot ground movement as small as four-tenths of an inch. The radar can "see" through clouds and darkness but not dense vegetation, ice, or snow. That means scientists can only use images of Cascade volcanoes taken in late summer.

The computer-driven technology has identified shifts in the ground at the Yellowstone caldera in Wyoming, the Long Valley caldera in California, Mount Etna in Italy, and at three Alaskan volcanoes. In addition to volcanoes, geologists are using INSAR to detect ground deformation related to earthquakes. The system allows the scientists to see how much change has taken place on the surface over a period of several years. The technology also can be used to examine changes in forests, glaciers, marshlands, and agricultural crops over a wide area.

peaks could also produce avalanches. The report also stated that during the past century at least five small lahars occurred on each of the mountains and nearby Broken Top volcano when warming glaciers unleashed a torrent of water. Fortunately the flows only swept across undeveloped areas within 6 miles of their sources.

More than a dozen glaciers cover portions along the flanks of the Three

South Sister (left) and North Sister rise above Black Butte Ranch in central Oregon.

Sisters. Collier Glacier, which stretches north-northwest from an altitude of about 9,000 feet between Middle Sister and North Sister, has been one of the most monitored glaciers in the Cascades. During the past century the 1.5-mile-long glacier has retreated more than a mile despite receiving about 30 feet of snowfall each winter. Scientists suspect global climate change may be involved.

Numerous volcanoes dot the area around the Three Sisters, with two especially notable nearby peaks--Mount Bachelor and Broken Top. The site of a popular ski resort, Mount Bachelor towers 9,065 feet over the landscape in an almost symmetrical cone. The young volcano is only about 15,000 years old. Broken Top, less than 5 miles southeast of South Sister, is 9,185 feet. Unlike the ice cream-cone structure of Bachelor, Broken Top is a craggy mountain with a large crater that was heavily eroded by glaciers during the last ice age. Like the other volcanoes in the Three Sisters area, its age is difficult to determine, but it is believed to be the oldest. Broken Top hasn't erupted in the last 10,000 years.

While the central Oregon Cascades continue to be a popular recreational area with climbers, hikers, and skiers drawn to their slopes, scientists are continuing to take measurements of the intriguing bulge west of South Sister. Instruments detected a swarm of more than 100 small earthquakes in the area in March 2004. They may be witnessing a typical Cascade occurrence of magma rising for a period, then receding. Or they could be seeing how a Cascade volcano is formed—perhaps the birth of a "West Sister."

For more information:

Deschutes National Forest

Sisters Ranger District
P.O. Box 249
Sisters, OR 97759
(541) 549–7700
www.fs.fed.us/r6/centraloregon

Bend–Fort Rock Ranger District

1230 Northeast Third Street, Suite A-262
Bend, OR 97701
(541) 383–4000
www.fs.fed.us/r6/willamette/recreation/tripplanning/wilderness/threesisters

USGS Cascades Volcano Observatory

vulcan.wr.usgs.gov/Volcanoes/Sisters

NEWBERRY VOLCANO

For the casual visitor to central Oregon's high desert region, Newberry Volcano is difficult to spot. The volcano is definitely a peculiar breed when compared to the majestic snowcapped Cascade peaks that crest 40 miles to the west.

Purists say it shouldn't be considered part of the Cascades; others say it's close enough. Regardless of its designation, Newberry Volcano is a unique formation that should not be overlooked. Unlike the steep-sided Cascade stratovolcanoes, Newberry is a gently sloping shield volcano that sprawls across 630 square miles of central Oregon just east of U.S. Highway 97, the area's main road. The volcano stretches about 34 miles north and south at its longest points and 22 miles east and west at its widest.

Labeling it a shield volcano, meaning it was formed primarily by flows of very fluid lava, is an oversimplification, however. Newberry is complex, having been shaped by a combination of lava flows and explosive eruptions during its 600,000-year history, with one ancient eruption covering the San Francisco Bay area 600 miles south with a half-inch-thick layer of ash.

The volcano has been best known for Newberry Crater, which sits within the 94-square-mile Newberry National Volcanic Monument south of the city of Bend. Technically the crater is actually a caldera, a huge depression left when eruptions eject such a large volume of magma from beneath a volcano that the ground collapses into the emptied space. Crater Lake, about 75 miles to the southwest, is one of the world's most impressive examples of a caldera.

The Big Obsidian Flow is a major feature in the Newberry caldera, which also has two lakes.

The 4-by-5-mile Newberry caldera, which contains two alpine lakes, is best viewed from 7,984-high Paulina Peak, which rises more than 3,500 feet above the surrounding area. Paulina and East Lakes may originally have been one body of water, but an eruption of lava and pumice split the caldera about 7,300 years ago.

Paulina Peak and Paulina Lake are named for a Snake Indian chief, while the volcanic area is named for a geologist and physician, Dr. John Strong Newberry, who took part in mapping future railroad routes through the area in 1855.

Newberry Crater's most impressive feature is called the Big Obsidian Flow. The volcano's last significant eruption 1,300 years ago produced the glassy black feature, which covers 1 square mile and is seventeen stories thick. A natural volcanic glass, obsidian is rich in silica and was used by Native Americans to fashion arrowheads, knives, and sharp tools. A half-mile interpretive trail through one corner of the flow gives visitors a close-up view of the formation.

THE OLDEST **HOUSE**

Insights into the earliest residents of the continent took a giant step forward at Newberry Volcano with the discovery in the 1990s of the oldest house ever found in western North America. The 9,400-year old house was likely buried in a volcanic eruption 7,700 years ago. Led by Thomas Connolly of the University of Oregon, the archaeologists who uncovered the house found a hearth and tools, pine support posts, and evidence of plants used as protective coverings. Connolly believes the house was likely a wickiup about 14 feet wide and 16½ feet long and may have been one of several domestic structures built for a family or several families. He says the site probably served as a staging area on Paulina Lake's shores for food gathering and hunting during the warm summer months.

By analyzing remnants from the hearth, the archaeologists concluded that the home's inhabitants ate hazelnuts, blackberries, chokeberries, and other fruits and nuts. Lodgepole and ponderosa pines were used for fuel, along with sagebrush wood. The sagebrush bark was used for rope, matting, and clothing. The early Oregonians also processed hardwood bark, bulrushes, and other plants to make coverings for the structure's floor and roof.

Blood residue found on tools indicated that rabbits may have been butchered in the house near the hearth, while bear, bison, sheep, deer, and elk were slaughtered outside the dwelling. Several of the house's perimeter posts were charred, indicating the structure had burned.

Research indicates the environment in Newberry Crater at the time was an open pine forest with a meadow and shrub-steppe understory. Studies found no evidence of human habitation in the caldera in the thousand years following the Mazama eruption. The house site is about 25 miles north of Fort Rock Cave, one of the Northwest's most famous archaeological sites. It was there in 1938 that anthropologists found seventy pairs of sandals made of sagebrush bark. When radiocarbon dating was later developed, one of the sandals was found to be about 9,000 years old.

While newcomers are flocking to central Oregon's high desert, Native Americans have long called the Newberry Volcano home. Several years ago archaeologists from the University of Oregon announced that they had discovered the oldest house ever found in western North America, right near the outlet of Paulina Lake in Newberry Crater. Radiocarbon dating showed the ancient structure to be about 9,400 years old. The remains of the house were buried beneath a layer of ash spewed by the massive eruption of Mount Mazama—the volcano that formed Crater Lake—about 7,600 years ago. Archaeologists uncovered a central fire hearth and tools, five lodgepole pine support posts, and plants used for roof and floor coverings.

Though Newberry shows no signs of erupting in the near future, geologists say the volcano is not extinct. They estimate hot magma lies only about 2 or 3 miles beneath Newberry Crater, where the next eruption probably will occur. A 1997 report by the Cascades Volcano Observatory said future eruptions likely would resemble those that have occurred at Newberry in the past 15,000 years, which have varied from highly explosive discharges of pumice and ash to quiet emissions of lava.

Eruptions could also emit plumes of ash and flows of mud and debris (lahars). The area most likely to be affected by lahars and floods would be the Paulina Creek valley, which drains from Paulina Lake through the west rim of Newberry Crater. The USGS says that should Newberry rumble back to life, lahars or floods from Paulina could reach the La Pine valley within thirty minutes. In recent years geothermal companies have been investigating an area just outside the Newberry National Volcanic Monument as a site for building a plant to produce electricity. Tests haven't found a commercially viable site, but explorations continue.

Established by Congress in 1990, the Newberry National Volcanic Monument protects the caldera area's distinctive geologic features. But the monument area covers only a small portion of the bigger volcano, which includes about 400 cinder cones—simpler volcanoes built primarily of particles of lava ejected from vents.

Newberry's unusual formations include Lava Butte, a cinder cone that rises 500 feet from its surroundings south of Bend. The Lava Lands Visitor Center on Lava Butte provides a wealth of information about the volcano and the national monument, as well as trails through lava beds and a young ponderosa forest.

About 1 mile south of Lava Butte on U.S. 97 is Lava River Cave, a mile-long lava tube that runs beneath the highway. Visitors can hike into a chilly forty-two-degree cave with lanterns and flashlights. Lava tubes are caves formed when streams of lava flow steadily in a channel for several hours or days. A solid crust, or roof, develops as the outer edges of the flow cools and remains when the hot interior lava has flowed out.

With its myriad striking features, Newberry Volcano is becoming a popular destination for visitors to central Oregon. At the same time, it is a volcano that scientists continue to monitor, as it could one day pose a hazard for the expanding resort communities that lie nearby.

For more information:

Deschutes National Forest
1645 Highway 20 E
Bend, OR 97701
(541) 383–5300
www.fs.fed.us/r6/centraloregon/newberrynvm

Ochoco National Forest
3160 Northeast Third Street
Prineville, OR 97754
(541) 416–6500

Crooked River National Grassland
813 Southwest Highway 97
Madras, OR 97741
(541) 475–9272

USGS Cascades Volcano Observatory
vulcan.wr.usgs.gov/Volcanoes/Newberry

CRATER LAKE

More than twenty-five million visitors have gazed on cobalt blue Crater Lake during the past century, its elegance unrivaled as it sits in the remnants of the Cascade volcano called Mount Mazama. Its 1,943-foot depth makes it the nation's deepest lake and the world's seventh deepest. One of the planet's clearest bodies of water, Crater Lake is a perfect picture of serenity. But its beauty belies its violent origins.

Mount Mazama began to form about a half-million years ago as magma rose from far below the earth's surface. The mountain's oldest rocks, about 420,000 years old, are visible at Mount Scott, which rises nearly 9,000 feet above sea level east of Crater Lake.

The Makalak people, ancestors of the present-day Klamath Indians, probably were the last to see 12,000-foot Mount Mazama intact before it exploded 7,700 years ago in the Cascades' largest known eruption. Living southeast of the mountain, they must have been terrified as repeated thunderous blasts spewed about 12 cubic miles of magma into the air in the form of ash and pumice. By comparison, the 1980 eruption of Mount St. Helens was paltry, removing only an estimated 0.12 cubic mile of magma. A layer of Mazama ash from the eruption can be found beneath the surface throughout Washington, Oregon, and western Canada.

Scientists who found Mazama ash in a core of ice in Greenland called the eruption one of the most "climatically significant" volcanic events in the Northern Hemisphere in the past 10,000 years. They estimate that the eruption filled the stratosphere with millions of tons of sulfuric acid droplets,

Wizard Island is a major feature on Crater Lake's West Side.

which blocked sunlight and might have lowered the earth's temperature by about one degree Fahrenheit.

Mount Mazama's massive eruption emptied its magma chamber about 3 miles below the surface, causing the top of the mountain to collapse on itself, forming a bowl-shaped depression called a caldera. The once majestic mountain lost 4,000 feet of its summit as it plunged inward.

Rain and snow then began accumulating in the 3,000-foot-deep caldera, but it took more than 200 years to form a lake. Volcanic activity continued, with the Wizard Island volcano rising 764 feet above the lake's surface in the caldera's western end. Another volcano, Merriam Cone, erupted underwater and now rises from the lake floor to within 500 feet of the surface near the volcano's north rim. An eruption 5,000 years ago formed a lava dome below the lake surface just east of Wizard Island, then eventually the volcano rested.

Fast-forward to June 12, 1853. John Wesley Hillman, Henry Klippel, and Isaac Skeeters are searching for the legendary Lost Cabin gold mine when they stumble upon the lake. Awed by its color, Skeeters suggests calling it "Deep Blue Lake." It's called "Lake Majesty" and "Blue Lake" by subsequent visitors over the next few years, before Oregon newspaper editor Jim Sutton dubs it "Crater Lake" after a trip to the caldera in 1869.

THE CRATER LAKE ECOSYSTEM

Crater Lake National Park contains more than just a spectacular body of water. The 183,000-acre park also boasts a rich treasure of ecological zones that are worth exploring.

Mount Mazama's eruption nearly 8,000 years ago erased much of the life in the vicinity, but the park demonstrates nature's resiliency after such a cataclysmic event. Researchers have learned a lot about animals and plants that have survived in ash-heavy soils. Several ecosystems reveal a variety of species that endure long winters and yet flourish during a short growing season.

The caldera's steep rim lies in a subalpine zone, where a few hardy species such as the peregrine falcon are able to thrive. Away from the caldera on the mountain's upper flanks, an abundance of fir trees provides habitat for small mammals. Clark's nutcrackers collect seeds from whitebark pines, storing some for winter food while burying others that emerge as seedlings a few miles away. At lower elevations Shasta red firs provide shade for mountain hemlock trees. The combination creates a complex old-growth forest that provides habitat for many species, such as the spotted owl and Pacific fisher.

One of the most intriguing areas is north of the caldera, where volcanic pumice and ash fill a mostly treeless basin. Only a few species can withstand this hostile environment of dry, nutrient-poor soils. In the summer grasses and tiny flowers emerge in green and pink clusters to provide a dash of color to the barren landscape. Lodgepole pines are mounting a slow invasion, which scientists expect to take thousands of years.

In 1885 an Oregon newcomer, William Gladstone Steel, was so inspired by the lake's beauty that he began a seventeen-year campaign to have the body of water and its surroundings protected. Steel formed a mountaineering group in Portland to promote the mountain and lake he wanted to save. His group, the Mazamas, gave the volcano its name. On May 22, 1902, Crater Lake became the nation's seventh national park.

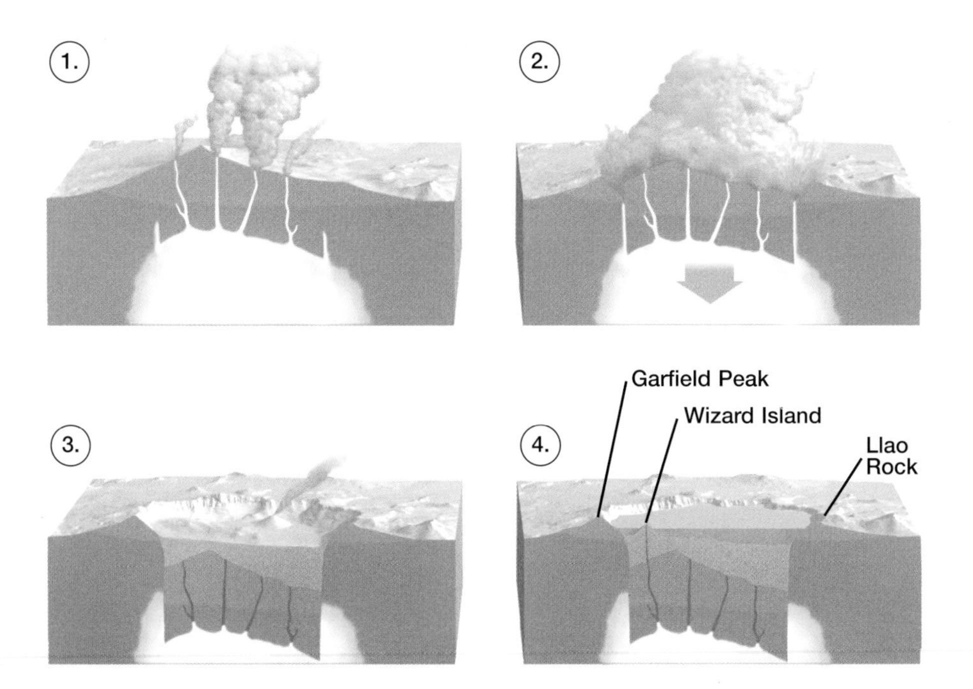

1. Mount Mazama grows over several hundred thousand years, with many eruptions. The vents erupt in intervals as short as a year or as long as thousands of years. The volcano rises to an elevation of 12,000 feet.
2. About 7,700 years ago, Mount Mazama erupts with a force forty times more powerful than the 1980 eruption of Mount St. Helens. A void is created beneath the volcano as its magma chamber is emptied.
3. The roof of the magma chamber collapses, forming the bowl-shaped depression known as a caldera. Weak explosions within the caldera follow.
4. Rain and snowfall begin filling the 3,000-foot-deep caldera with water, forming a nearly 2,000-foot-deep lake over a period of about 300 years. Volcanic activity continues, forming Wizard Island in the lake's west side.

Crater Lake's location within an ancient volcano provides it with several unique characteristics. Abundant snowfall, averaging nearly 45 feet each year, supplies the lake with almost all its water. As the lake has no surface outlets, evaporation and underground seepage balance the input of precipitation and keep the lake at a stable level.

Because Crater Lake is filled primarily by snowfall, it's one of the purest lakes in the world. The lake's striking blue color is the result of its crystal-clear water, with sunlight able to penetrate deep down. Water molecules absorb the longer wavelengths of red, orange, yellow, and green light, while shorter wavelengths of blue light scatter and shoot back up to the surface.

The lake holds nearly five trillion gallons of water, enough to give every American about forty-five gallons of water a day for a year. The protected lake's pristine condition and its remote location in southern Oregon make it a natural laboratory for scientists to study and monitor changes in the environment, both locally and globally.

Scientists have long been lured to the lake. An 1896 article in *Scientific American* magazine called Crater Lake "the most remarkable body of water on the continent, whether regarded as scenery or as an object of scientific interest." In the 1880s the U.S. Geological Survey conducted the first systematic study of the lake, mapping its bottom by lowering piano wire and lead weights in more than a hundred sites. Using slightly more sophisticated technology, USGS scientists in the year 2000 used a high-tech sonar mapping system to take sixteen million soundings of the lake, establishing the depth at 1,943 feet. They also used the data to produce colorful computer-generated maps of the lake floor and three-dimensional views of the caldera's underwater structures.

In 1988 and 1989 oceanographers Jack Dymond and Robert Collier of Oregon State University were the first people to get a direct look at the caldera floor when they used a one-person submersible to explore the lake. They found temperature and chemical evidence that suggests hot springs exist on the lake floor but did not see fluids flowing from any vents. On one dive Collier found 30-foot-tall spire formations in the northeast part of the lake, direct evidence that hydrothermal processes have occurred on the lake floor.

National Park Service biologist Mark Buktenica also used the small submarine to explore biological features on the lake floor. He found numerous plants and animals living in the lake, including nematodes, earthworms, midge flies, and other organisms. Though fish are not native to the lake, a few thousand rainbow trout and kokanee salmon flourish, the result of stocking between 1888 and 1942.

The National Park Service began a monitoring program in 1982 to keep an eye on Crater Lake's remarkable features. Scientists have found that the lake's clarity varies from year to year and even seasonally. Storms and erosion from the steep cliffs add material to the lake that can affect its clearness. In 1997 researchers recorded a clarity reading of 142 feet, a world record.

THE LAKE'S **FLOOR**

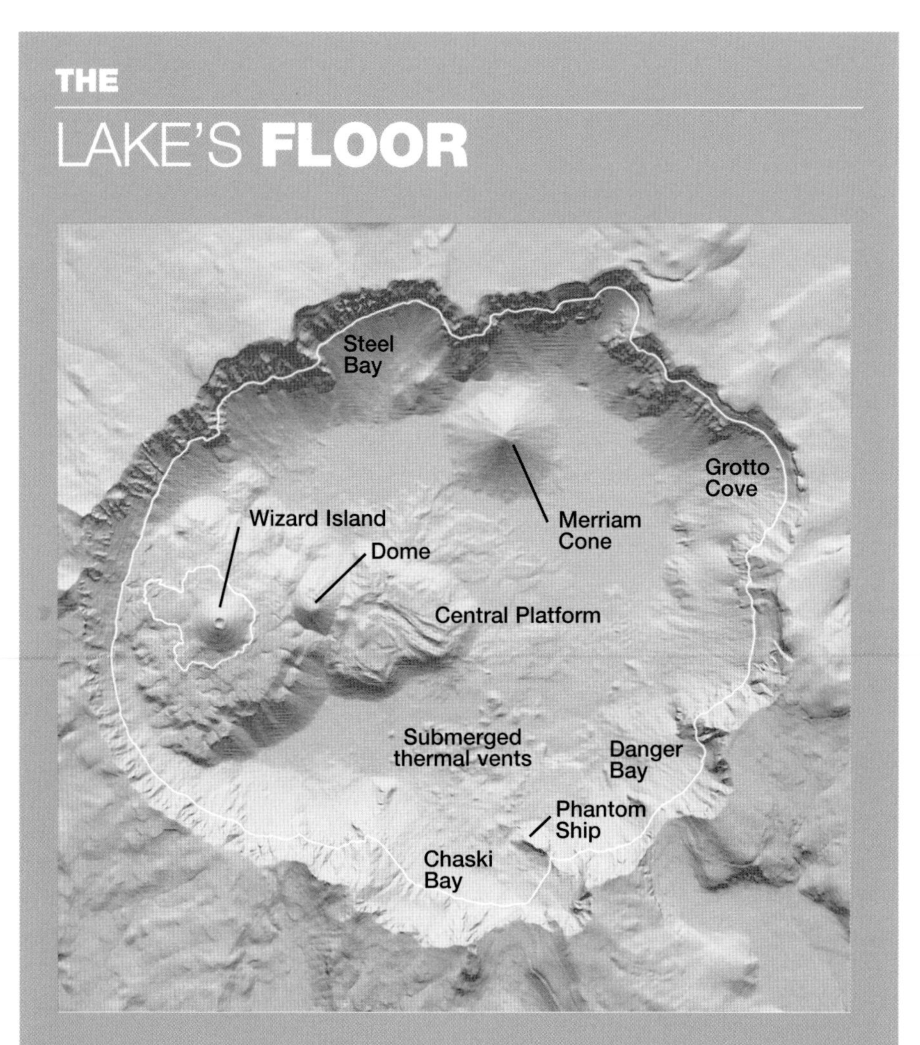

In August 2000 scientists made a detailed map of the caldera floor using a sophisticated sonar system. The image shows Wizard Island at left and Merriam Cone at the top center, which rises from the lake floor to within 500 feet of the lake's surface.

Thanks to sophisticated high-tech equipment, humans are getting their first good look at the rugged volcanic terrain that lies beneath Crater Lake's glistening blue surface. Because the lake is so deep, it's not easy to explore even with small submarines since light does not penetrate to the bottom of the lake.

But state-of-the art sonar technology has easily pierced through that immense body of water to provide a detailed map of the caldera floor. In August 2000 scientists with the U.S. Geological Survey completed a detailed mapping of the lake floor. The work produced colorful three-dimensional images that scientists can use to get a better understanding of the volcano's geologic history. The project took five days and provided impressive details of many geologic features, including the Chaski landslide. That massive avalanche spilled down the volcano's fragile south rim shortly after the cataclysmic eruption 7,700 years ago and before the lake had fully formed an estimated 250 years later. The crew used a high-resolution, multi-beam sonar-mapping system mounted on a 26-foot research vessel, the *Surf Surveyor.*

The sonar instruments took sixteen million soundings of the lake floor, providing data for a much more detailed map than had ever been made. The technology can generate topographical maps accurate to within 1½ feet and can detect objects larger than 3 feet in diameter. In addition to depth data, the system also records "backscatter," the strength of sound energy that bounces back. With that information the scientists can identify sand, mud, or other materials that make up the lake floor. The work also resulted in a new depth reading for the nation's deepest lake. Instead of 1,932 feet deep, the more precise sonar data tacked on another 11 feet, making it 1,943 feet.

Given its history, Mount Mazama probably will reawaken someday, scientists say, but they are observing no present signs of activity. A report by the USGS in 1997 said the western part of the caldera is considered the most likely site of volcanic activity. One potential hazard is an earthquake-caused landslide from the steep cliff walls above the lake. Landslides also could cause large waves on the lake that could pose a hazard for tour boats. But given the low level of earthquake activity in and around the park, it's not considered a significant threat.

There's plenty for visitors to see in the park, which is accessible after the snow melts early in the summer. Tour boats provide visitors with close-up looks at several features, including Wizard Island and the Phantom Ship, a 400,000-year-old lava structure in the lake's south side. High above the lake surface, which sits at 6,176 feet above sea level, is the 33-mile rim drive around the body of water. Crater Lake Lodge, which opened on the lake's south rim in 1915, underwent a $15 million remodeling in the early 1990s. All told, 500,000 visitors flock to take in the beauty of Crater Lake each year. It's a sight few people forget.

For more information:

Crater Lake National Park
P.O. Box 7
Crater Lake, OR 97604
(541) 594–3100
www.nps.gov/crla

USGS Cascades Volcano Observatory
vulcan.wr.usgs.gov/Volcanoes/CraterLake

Books
Crater Lake National Park: A History by Rick Harmon (Oregon State University Press, 2003)

MOUNT MCLOUGHLIN

Mount McLoughlin is southern Oregon's signature peak, dominating the skyline about 30 miles south of Crater Lake and 70 miles north of Mount Shasta. Midway between the cities of Klamath Falls and Medford, the steep-sided stratovolcano rises to a lofty 9,495 feet. The mountain provides an impressive landmark that can be viewed from the Rogue Valley along Interstate 5 to the west and from the Klamath Basin along U.S. Highway 97 to the east.

The cone of the volcano is only about 100,000 years old, with its last major eruption occurring about 20,000 years ago. Although it looks symmetrical from some perspectives, ice-age glaciers cut away most of the volcano's upper northeast side.

Settlers in the nineteenth century called it Mount Pit because of pits dug by Indians to trap game. The Oregon legislature named it Mount McLoughlin in 1905 to honor the "father of Oregon," John McLoughlin, who served as the chief factor of the Hudson's Bay Company at Fort Vancouver.

A challenging 11-mile round-trip trail to the volcano's summit offers summer visitors spectacular views of the surrounding countryside.

The view from Mount McLoughlin, the highest Cascade peak between the Three Sisters volcanoes to the north and Mount Shasta to the south, includes the notable nearby volcanoes of Brown Mountain and Pelican Butte. Brown Mountain, about five miles southeast of Mount McLoughlin, is a 7,300-foot-high shield volcano that produced an extensive lava field about 15,000 to 20,000 years ago. Pelican Butte, another broad shield volcano, rises 8,036

Mount McLoughlin's spectacular peak towers over southern Oregon.

feet above sea level about 5 miles northwest of Mount McLoughlin. The mountain, which exceeds McLoughlin in volume, has near-perfect symmetry that is blemished only by two large bowls on its northeast side, cut by ice-age glaciers.

Mount McLoughlin is adjacent to the 40,000-acre Klamath Marsh National Wildlife Refuge, which is home to a variety of bird species, including the American bald eagle and the American white pelican. A narrow, rough road leads to the summit, which provides a panoramic view of the area's numerous volcanoes and lakes.

For more information:

Fremont-Winema National Forests
2819 Dahlia Street
Klamath Falls, OR 97601
(541) 883–6714
www.fs.fed.us/r6/winema

USGS Cascades Volcano Observatory
vulcan.wr.usgs.gov/Volcanoes/McLoughlin

MOUNT SHASTA

From his first look at majestic Mount Shasta in 1874, naturalist John Muir knew he had stumbled upon one of America's most awe-inspiring scenes. "When I first caught sight of it over the braided folds of the Sacramento Valley, I was fifty miles away and afoot, alone and weary. Yet all my blood turned to wine, and I have not been weary since," Muir wrote.

Anchoring the southern edge of the Cascades about 40 miles south of the Oregon-California border, snow- and glacier-capped Mount Shasta can be seen from up to 100 miles in all directions and serves as an inspiring landmark to travelers along busy Interstate 5. Though it's highly visible, Mount Shasta sits in a relatively remote area far from any major cities. The mountain is included in the Shasta-Trinity National Forest and is part of the 38,200-acre Mount Shasta Wilderness Area established by Congress in 1984.

The Cascades' second highest mountain towers over the countryside at 14,162 feet, only about 250 feet lower than Mount Rainier. With a surface mass of more than 80 cubic miles, Shasta is the most voluminous of the Cascade stratovolcanoes and ranks as one of the world's largest steep-sided stratovolcanoes. Nearby Medicine Lake Volcano, a shield volcano similar to the Hawaii volcanoes, is the most voluminous volcanic center of the Cascade Range.

Some accounts say that the French explorer Jean-Francois de Galaup de Laperouse spotted the volcano erupting as he sailed off the Pacific coast in 1786, but skeptics think he may have been seeing smoke from Indian fires rather than an eruption. Regardless, scientists believe the volcano last erupted about 200 years ago and undoubtedly will erupt again. It's potentially the

Mount Shasta is the Cascades' second highest peak.

most dangerous volcano in northern California and—other than Mount St. Helens—has produced more documented eruptions in the past 4,000 years than any other Cascade volcano.

Native Americans had lived in the area for several millennia, calling their revered mountain by a variety of names. Historians are uncertain about the present name's origin, but it might have been the name of an Indian tribe in the region.

Fur trapper Peter Skene Ogden is credited for giving the mountain an approximation of its current name. He wrote in his journal in 1827 that he had seen a mountain "equal in height to Mount Hood" that he named "Mt. Sastise." Some historians believe the mountain he described was Mount McLoughlin in southern Oregon rather than Mount Shasta, and a later map is believed to have mistakenly placed Ogden's Sastise mountain where Shasta is. The modern name first appeared in 1850 when the California legislature named the surrounding county "Shasta."

Mount Shasta began forming at least 600,000 years ago as magma from deep below the surface in the hot Juan de Fuca Plate found its way upward. One of the Shasta region's most unique features isn't on the mountain itself, but is an area of the Shasta Valley northwest of the peak that is filled with hundreds of hills, mounds, and ridges. Geologists were puzzled by the hummocky scene until the massive landslide during the 1980 Mount St. Helens eruption provided the key clue that helped solve the mystery. What scientists realized is that a huge avalanche destroyed much of ancient Mount Shasta about 300,000 years ago. The avalanche swept more than 30 miles from the present volcano's summit, blanketing about 250 square miles of the western part of the Shasta Valley under 11 cubic miles of material. The landslide dumped over twenty times more material into Shasta Valley than Mount St. Helens released in its devastating avalanche.

Scientists are not certain what caused the enormous slide. There's no evidence of an eruption, but any signs of a blast would have eroded away long ago. One theory is that the avalanche began with a series of large landslides of water-laden rock that carved into the mountain.

While the avalanche feature makes Mount Shasta unique among Cascade volcanoes, the mountain also has another characteristic that makes it unusual. Eruptive episodes have built four cones on the volcano, each providing a tale about Shasta's ancient history:

- Sargents Ridge cone rises on the volcano's high southwest flank, the eroded, ragged remnant of a series of eruptions that occurred about 100,000 to 200,000 years ago.
- Misery Hill cone on the north side rose about 30,000 to 50,000 years ago, providing much of the material for the present volcano's upper portion.
- Shastina cone is a distinctive volcanic feature that stands alone on the mountain's west side. The cratered cone was formed about 9,500 years ago. Low on Shastina's southwestern flank and towering above I–5 is a 2,500-foot dome called Black Butte, which formed during eruptions that occurred about the same time as the Shastina-forming events.
- Hotlum cone, which forms the volcano's summit and the north and northwest slopes, is the energetic youngster of the mountain. Though it probably began forming about the time the Shastina cone arose, it's mostly been growing in the past 6,000 years. It has been the center of activity the past few thousand years, last erupting about 200 years ago.

Seven major glaciers adorn Mount Shasta: Bolam, Hotlum, Konwakiton, Mud Creek, Whitney, Wintun, and Watkins. Whitney is California's longest glacier, a river of ice flowing more than 2 miles down Shasta's northwest

A massive landslide left hundreds of hummocks near Mount Shasta.

flank in the valley between the Hotlum and Shastina cones.

Sulfur-spouting fumaroles and boiling springs high on the mountain are an indication that hot magma still resides deep beneath the mountain and serve as a signal that the volcano isn't dead. Future eruptions would certainly cause the rapid melting of glacial ice as well as snowpack, resulting in large debris-filled mudflows, or lahars, down the mountain's slopes and endangering the towns of Weed, McCloud, Mount Shasta, and Dunsmuir at the volcano's base.

Lahars aren't the only danger. As in the past, eruptions could cause pyroclastic flows, ground-hugging avalanches of extremely hot ash, pumice, rock fragments, and volcanic gas that rush down the side of a volcano as fast as 65 mph or more.

Geologists with the Cascades Volcano Observatory point out that Mount Shasta has erupted on average at least once every 600 years in the past 4,500 years, and it's erupted at least three times in the past 750 years. They say there's no regular interval between eruptions, but there's no reason to think Shasta won't erupt again.

For more information:

Shasta-Trinity National Forest
2400 Washington Avenue
Redding, CA 96001
(530) 244–2978
www.fs.fed.us/r5/shastatrinity

Mount Shasta Companion
College of the Siskiyous
800 College Avenue
Weed, CA 96094
(530) 938-5555
www.siskiyous.edu/Shasta

USGS Cascades Volcano Observatory
vulcan.wr.usgs.gov/Volcanoes/Shasta

USGS Volcano Hazards Program
volcanoes.usgs.gov/Hazards/Where/ShastaDanger

MEDICINE LAKE VOLCANO AND LAVA BEDS NATIONAL MONUMENT

Medicine Lake Volcano is perhaps the least known of the thirteen Cascade volcanic centers, residing in a sparsely populated high-desert area of north-central California. Unlike the classic steep-sided stratovolcano structure of Mount Shasta only 30 miles to the southeast, Medicine Lake is a broad shield volcano that sprawls across the landscape.

But don't sell this volcano short. Medicine Lake ranks as the largest of the Cascade volcanoes in volume, with its low-profile structure covering nearly 800 square miles and composed of more than 145 cubic miles of material. The volume is ten times that of Mount Shasta, one of the world's largest stratovolcanoes.

Medicine Lake is often compared to Newberry Volcano, a shield volcano that covers 630 square miles west of the Cascade crest in central Oregon. Each features a caldera, or depression, on its summit caused by the inward collapse of the volcano after eruptions emptied the magma chambers far below the surface.

The highest point in Newberry's caldera is Paulina Peak at 7,984 feet; the highest point in Medicine Lake's caldera is Mount Hoffman, which is perched on its north rim at 7,913 feet. The flanks of each volcano are blanketed with hundreds of cinder cones and lava flows. Each sports a large flow of obsidian, a picturesque lake, and lava tubes.

Medicine Lake Volcano is roughly a half-million years old, with most of its current structure forming in the past 40,000 years. Eruptions have caused its peak to collapse several times, which has resulted in the 4.5-by-7.5-mile

Medicine Lake is the key feature of the Medicine Lake caldera. Mount Shasta is on the horizon.

caldera at its summit. Its namesake lake, which covers more than 0.6 square mile and is 152 feet deep at its deepest point, sits within the caldera at an elevation of 6,680 feet.

The U.S. Geological Survey estimates that the volcano has erupted at least seventeen times in the past 12,000 years ago. Eight eruptions shook the mountain over a period of just a few centuries about 10,500 years ago, then the volcano rested for more than 6,000 years before coming back to life. Another restless period began about 3,000 years ago, with the volcano erupting eight more times before its last event in about A.D. 1100.

The last explosion produced the massive Glass Mountain Obsidian Flow on the volcano's east flank. The 3-mile-long flow covers 570 acres with stony gray dacite obsidian. Native Americans used Glass Mountain as an obsidian quarry, with the black rock prized for making sharp-edged tools and as a popular trade item.

Several tribes lived in the area, including the Modoc people, who had a summer village in the Medicine Lake basin. With the rugged volcanic terrain providing a natural lava fortress, the tribe was able to hold off U.S. troops for more than five months in an area called Captain Jack's Stronghold during the Modoc War of 1872–73.

A 73-square-mile portion of Medicine Lake Volcano is set aside as the Lava Beds National Monument. The federally protected area, designated a national monument in 1925 and managed by the National Park Service, is on the volcano's northeast flank adjacent to Tule Lake's southern shore.

With features similar to Idaho's Craters of the Moon National Monument and Hawaii Volcanoes National Park, the Lava Beds monument provides visitors with a look at a variety of volcanic features. About 70 percent of the monument is blanketed with material from thirty separate lava flows from Medicine Lake Volcano. In addition, the park includes rock art up to nearly 1,500 years old left by the region's original inhabitants.

The park is perhaps best known for its 435 lava tubes, the highest concentration of such features in North America. Most of the caves formed during a Medicine Lake eruptive period about 30,000 years ago. The tubes are created when hot lava flows from a volcano at about 1,800 degrees. The flow's surface cools rapidly and begins to harden, acting as insulation for the fast-moving lava beneath it. When the lava drains, the outer shell is left, forming a tunnel. Formations called "lavacicles," produced when molten lava drips and splashes, hang from the ceilings in the tunnels.

Several of the caves bear a variety of colorful names, including Mushpot, Thunderbolt, Hopkins Chocolate, Golden Dome, Sunshine, Blue Grotto, Valentine, Skull, and Hercules Leg. Catacombs Cave, at 1.3 miles, is the longest single cave.

All of the features are from the distant past at Medicine Lake, which has shown no signs of eruptive activity in the past 1,000 years. However, geologists say that it likely will erupt again. Looking at its past, they predict that an eruption would produce small pyroclastic flows, along with emissions of ash and pumice. Future volcanic activity also may generate lava flows and form cinder cones. Because of Medicine Lake's remote location, an eruption would pose a relatively small threat to people, though it could hinder traffic on nearby I–5, the major artery between California and the Pacific Northwest.

Its remote location, however, is what makes Medicine Lake a popular place to escape the hustle and bustle of urban life and see what nature's dramatic processes have produced.

The Glass Mountain Obsidian Flow covers 570 acres. Mount Shasta is in the distance.

For more information:

Lava Beds National Monument

1 Indian Well Headquarters
Tulelake, CA 96134
Headquarters: (530) 667–2282
Visitor information: (530) 667–2282, ext. 232
www.nps.gov/labe

Modoc National Forest

800 West Twelth Street
Alturas, CA 96101
(530) 233–5811
www.fs.fed.us/r5/modoc/recreation/modocscenicbyway.shtml

USGS Cascade Volcano Observatory

Vulcan.wr.usgs.gov/Volcanoes/MedicineLake

LASSEN PEAK

Memorial Day 1914 became an especially memorable one for the few people living near Lassen Peak in sparsely populated north-central California. After 27,000 years of slumber, the volcano jolted awake with a steam explosion from a new vent near its summit.

That small burst was only a hint of what was to come from the Cascades' southernmost volcano. Within a year more than 180 steam explosions caused by pressurized magma-heated water had carved a 1,000-foot-wide crater on the mountain.

But almost a year later, on May 14, 1915, the volcano changed tactics. Rather than just emitting powerful steam that spewed plumes of smoke, ash, and large rocks, the mountain began throwing blocks of lava down its flanks. By the following morning a dome of lava had filled the crater, preparing the peak for even more violent events.

On May 19 a large steam explosion shattered the lava dome, causing hot lava to collapse on the peak's snow-covered upper flanks. The heated rocks unleashed a half-mile-wide avalanche 4 miles down the mountain's steep northeast side. Snow melted by the hot rocks resulted in a lahar that sped 7 miles down Lost Creek. Both the avalanche and lahar flooded the lower Hat Creek Valley during the early morning hours of May 20, destroying six houses. The few people who were there escaped with only minor injuries.

Still, the volcano was far from finished. At 4:45 P.M. on May 22, Lassen Peak exploded in a huge eruption. The blast sent a plume of ash and gas more than 40,000 feet into the atmosphere and produced a hot mixture of

Lassen Peak, which erupted in 1915, is now part of a national park.

ash, rock, and gas called a pyroclastic flow down the volcano, destroying a 3-square-mile area. Ash fell as far as Winnemucca, Nevada, 200 miles to the east. No one was injured in the blast because the previous twelve months of activity had put residents on high alert. Lassen Peak continued to rumble for two more years, but nothing came close to the 1915 event. It was that eruption that first drew widespread attention to the dangers of Cascade volcanoes. The spectacular blast and the rugged volcanic scenery also led to the area being designated a national park in 1916. The last eruption at the peak occurred in 1921.

The activity in the early 1900s was just the latest in this volcanic center's long history. About 600,000 years ago eruptions began building a large volcano known as Mount Tehama or Brokeoff Volcano. The volcano became inactive about 400,000 years ago, and its remnants can now be seen at Brokeoff Mountain, Diamond Peak, Mount Diller, and Mount Conard.

Eruptions continued throughout the area, however, with more than thirty mounds of lava being built. Lassen Peak, one of the planet's largest lava domes at 10,457 feet, formed in an eruptive period about 27,000 years ago. The next eruptive period occurred about 1,100 years ago in an area northwest of Lassen Peak. There eruptions formed a cluster of lava domes called the Chaos Crags, one of the national park's most distinctive features.

About 300 years ago a series of large avalanches swept down the north side of Chaos Crags at speeds of more than 100 mph. The avalanches formed Chaos Jumbles, an aptly named pile of rubble that blankets a 2.5-square-mile area.

Although Lassen Peak has been quiet since its last eruptions, steam rising from small vents serves as a reminder that the volcano is not dead. The 106,366-acre national park also includes several hot springs and steaming fumaroles, including colorfully named Bumpass Hell, Sulphur Works, Little Hot Springs Valley, Cold Boiling Lake, and Devil's Kitchen Valley. The temperature at each fumarole usually is close to the boiling point for its particular altitude. For example, Bumpass Hell south of the summit has a temperature of 198 degrees. Temperatures as high as 230 degrees have been recorded in the park's springs.

Because Lassen Peak is the only Cascade volcano other than Mount St. Helens to have erupted during the past century, scientists with the U.S. Geological Survey are keeping a close watch on the mountain. They say future eruptions could produce pyroclastic flows, small- to moderate-size debris flows, cinder cones, and lava flows.

The volcano is named for Peter Lassen, a native of Denmark who opened the Lassen Emigrant Trail in 1848 when he led a twelve-wagon train from Missouri to California. The trail was used extensively until 1853 during the California gold rush.

Lassen Volcanic National Park is just a small part of the Lassen National Forest, which covers 1,875 square miles and is part of seven counties. The forest lies at the heart of what is called the Crossroads, where the Sierra Nevada Range, the Cascade Range, the Modoc Plateau, and the Great Basin meet.

Snow covers much of the park from late October through early June, making for a short summer visitor season. Although Lassen Peak is off California's well-traveled tourist trail, its diversity of volcanic features makes it well worth a trip.

For more information:

Lassen Volcanic National Park
P.O. Box 100
Mineral, CA 96063
(530) 595–4444
www.nps.gov/lavo

Lassen National Forest
2550 Riverside Drive
Susanville, CA 96130
(530) 257-2151
www.fs.fed.us/r5/lassen

USGS Cascade Volcano Observatory
vulcan.wr.usgs.gov/Volcanoes/Lassen

INDEX

PHOTOGRAPHIC CREDITS

The author and editors greatly appreciate the assistance of the United States Geological Survey and the Cascade Volcano Observatory for providing the photographs for this book.

Title page: Lyn Topinka, USGS/CVO; Lyn Topinka USGS/CVO; Harry Glicken, USGS/CVO

1: Lyn Topinka, USGS/CVO; Lyn Topinka, USGS/CVO; Harry Glicken, USGS/CVO;

2: Lyn Topinka, USGS/CVO

9: Thomas J. Casadevall, USGS/CVO; Daniel Dzurisin, USGS/CVO; Lyn Topinka, USGS/CVO

10: Peter W. Lipman, USGS/CVO

12: Daniel Dzurisin, USGS/CVO

14: Lyn Topinka, USGS/CVO

17: Thomas J. Casadevall, USGS/CVO; Wade B. Clark Jr., Mount Baker-Snoqualmie National Forest; Bob Symonds, USGS/CVO

18: Thomas J. Casadevall, USGS/CVO

21: Terry Richard, D. Mullineaux, USGS/CVO; Gary Paull, Mount Baker-Snoqualmie National Forest

22: Terry Richard

25: David E. Wieprecht, USGS/CVO; Lyn Topinka, USGS/CVO; David E. Wieprecht, USGS/CVO

26: David E. Wieprecht, USGS/CVO

28: David E. Weiprecht, USGS/CVO

31: Donald A. Swanson, USGS/CVO; Peter W. Lipman, USGS/CVO; Donald A. Swanson, USGS/CVO

32: Donald A. Swanson, USGS/CVO

34: Donald A. Swanson, USGS/CVO

35: Lyn Topinka, USGS/CVO

37: Lyn Topinka, USGS/CVO

38: Lyn Topinka, USGS/CVO

41: Lyn Topinka, USGS/CVO; Lyn Topinka, USGS/CVO; Lyn Topinka, USGS/CVO

42: Lyn Topinka, USGS/CVO

43: Lyn Topinka, USGS/CVO

45: Lyn Topinka, USGS/CVO; Cynthia Gardner, USGS/CVO; David E. Wieprecht, USGS/CVO

46: Lyn Topinka, USGS/CVO

49: David E. Wieprecht, USGS/CVO, Lyn Topinka, USGS/CVO; Dan Miller, USGS/CVO

50: David E. Wieprecht, USGS/CVO

53: David E. Wieprecht, USGS/CVO; David E. Wieprecht, USGS/CVO; Lyn Topinka, USGS/CVO

54: David E. Wieprecht, USGS/CVO

56: Richard L. Hill

59: Lyn Topinka, USGS/CVO; Lyn Topinka, USGS/CVO; Robert A. Jensen, Newberry National Volcanic Monument

60: Lyn Topinka, USGS/CVO

65: Lyn Topinka, USGS/CVO; W. E. Scott, USGS/CVO; Connie Hoong, USGS/CVO

66: Lyn Topinka, USGS/CVO

73: Terry Richard, USGS/CVO; Lyn Topinka, USGS/CVO; C. D. Miller and D. Mullineaux, USGS/CVO

74: Terry Richard

75: David E. Wieprecht, USGS/CVO; Harry Glicken, USGS/CVO; Lyn Topinka, USGS/CVO

76: David E. Wieprecht, USGS/CVO

78: Harry Glicken, USGS/CVO

81: Julie Donnelly-Nolan, USGS/Menlo Park, Julie Donnelly-Nolan, USGS/Menlo Park, Julie Donnelly-Nolan, USGS/Menlo Park

82: Julie Donnelly-Nolan, USGS/Menlo Park

84: Julie Donnelly-Nolan, USGS/Menlo Park

85: David E. Wieprecht, USGS/CVO; Lyn Topinka, USGS/CVO; Lyn Topinka, USGS/CVO

86: David E. Wieprecht, USGS/CVO

ABOUT THE **AUTHOR**

Richard L. Hill has been the science writer for ***The Oregonian*** since 1988. He has received several honors for his work, including the American Geophysical Union's David Perlman Award for Excellence in Science Journalism and the C.B. Blethen Memorial Award for Distinguished Reporting. He serves on the board of the National Association of Science Writers and holds a journalism degree from the University of Texas at Austin. He lives in the Portland, Oregon, area with his wife Tracey and three children.

ABOUT THE ILLUSTRATOR

Steve Cowden has worked for ***The Oregonian*** for fourteen years creating award winning informational graphics. He has also illustrated many textbooks and a children's picture book. He lives in Lake Oswego, Oregon, with his wife Karen and four children.